82■00124

Concrete Flatwork Manual

By George Gilson

D0813722

Craftsman Book Company
6058 Corte Del Cedro, Box 6500,
Carlsbad, CA 92008

This book is dedicated to:
The Gilson, Homenick, Oliveri
Potter, and Stahl families

Library of Congress Cataloging in Publication Data

Gilson, George.
    Concrete flatwork manual.

    Includes index.
    1. Concrete construction. I. Title.
TA682.4.G54   624.1'834
ISBN 0-910460-93-0

82-1398
AACR2

Edited by Lois Larson
Illustrations by Bill Lahea

Contents

1

Fundamentals of Concrete

A nyone working with concrete should understand its general characteristics, properties and tendencies to guarantee correct handling from start to finish. Concrete is a plastic material consisting of sand, crushed rock or gravel, water and cement. The resulting material can be shaped and formed into practically any design you require. Concrete hardens in a process called *hydration* in which the water in the mixture combines chemically with the other components to produce a hard, rocklike conglomerate. The properties of *durability, plasticity, watertightness, strength* and *workability* make concrete a desirable and enduring substance for interior or exterior use.

Strength is measured in psi or pounds per square inch, which indicates the compressive stress tolerances of a given batch of concrete. To fully comprehend psi, you must first understand the batch formulations of concrete. The two basic mixes are 3,000 psi concrete or five-bag mix, and 4,000 psi or six-bag mix. The batch formulation by volume for five-bag mix is 1:2¾:4, which means concrete to be mixed will consist of 1 part cement with 2¾ parts sand and 4 parts stone. The batch formula for six-bag-mix concrete is 1:2¼:3, which indicates concrete to be mixed will consist of 1 part cement with 2¼ parts sand and 3 parts stone.

Generally, the higher the cement ratio in the mix, the greater the psi tolerance. A cubic yard of concrete with a 1:2¼:3 mix requires 6 bags of cement, 14 cubic feet of sand, and 18 cubic feet of stone. When ordering concrete for most projects, a six-bag-mix with a 4,000 psi compressive strength rating is usually ideal.

The quality of concrete can be undermined by an excessive quantity of water; the more water added to a mix, the weaker it becomes. Quality concrete differs in the amount of water it requires; a variance known as *slump*. Slump affects workability because the lower the slump, the less water there is in the mix, and the harder it is to place the cement properly.

You can increase durability of the concrete mixture by adding an air entraining agent, which causes tiny bubbles of air to be dispersed in the mix. This greatly improves wear resistance to alkali and salt, which is important in exterior concrete work.

Entrained air should be used in all concrete exposed to freezing and thawing and may be used for mild exposure conditions to improve workability. It is recommended for all paving concrete regardless of climate. The recommended total air contents for air-entrained concretes are shown in Table 1-1. When mixing water is held constant, the entrainment of air will increase slump. When cement content and slump are held constant, less mixing water is required. Less water helps to offset possible strength decreases and results in improvements in other paste properties such as permeability. The strength of air-entrained concrete may be equal, or nearly equal, to non-air-entrained concrete when their cement contents and slumps are the same.

Maximum Size of Aggregate in Inches	Recommended Average Total Air Content, %	Gallons of Water Per Cubic Yard		
		1" to 2" Slump	3" to 4" Slump	5" to 6" Slump
3/8	7.5	37	41	43
1/2	7.5	36	39	41
3/4	6.0	33	36	38
1	6.0	31	34	36
1-1/2	5.0	29	32	34
2	5.0	27	30	32
3	4.0	25	28	30
6	3.0	22	24	26

Recommended air and water content for air-entrained concretes
Table 1-1

Type of construction	Slump, inches	
	Maximum	Minimum
Reinforced foundation walls and footings	6	3
Unreinforced footings, caissons, and substructure walls	4	1
Reinforced slabs, beams, and walls	6	3
Building columns	6	4
Pavements	3	1
Heavy mass construction	3	1
Bridge decks	4	3
Sidewalk, driveway, and slabs on ground	6	3

Recommended slumps for various types of construction
Table 1-2

A slump test is generally used to measure the consistency of concrete. It should not be used to compare mixes with wholly different proportions or mixes with different kinds or sizes of aggregates. When used to test different batches of the same mixture, changes in slump indicate changes in materials, mix proportions, or water content. Acceptable slump ranges are indicated in Table 1-2.

For most projects, a 5-inch slump usually allows the dispatcher enough placement time, but advice from the local concrete company can certainly be well taken.

2
Tools

Many tools are required in concrete construction. The following list describes the basic tools and explains their uses in the field.

Level: Many contractors consider the level one of their most important and useful tools. No tool is more widely used in curb and gutter work. A quality level is made of wood with brass edging. This type of level ranges from 3 to 6 feet in length and has three gauges: two for horizontal leveling and one for vertical plumbing as shown in Figure 2-1.

Level transit: Shown in Figure 2-2, the level transit is a precise instrument for determining elevations for setting forms. Place the transit legs into the ground and screw the instrument onto the top faceplate with a threaded female connector. Then swivel the transit body directly across two leveling screws and turn it until the bubble in the leveling vial of the transit is centered between two

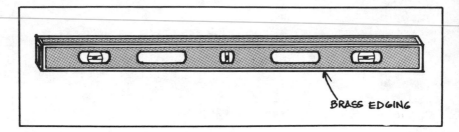

BRASS EDGING

Wood level
Figure 2-1

parentheses marks. These two marks indicate that the transit is in the level position. Rotate the instrument across the remaining two leveling screws, stop, and then adjust until the transit vial bubble is in the level position. Look through the instrument at a wood ruler held by another person on any point desired. The instrument, which sites along a fixed line of the level, will indicate an elevation on the ruler when focused properly. If the desired point of elevation is known, vertical adjustments for framing can be achieved more easily. The reading on the ruler as seen through cross hairs in the instrument indicates the elevation of that particular ground point.

Long-handle concrete placer: This tool, shown in Figure 2-3 is commonly known as a *come-along*. The flat end of the tool is used to pull and place concrete. Its long handle makes it much easier to pull concrete. It is also an extremely effective device for accurately grading concrete flatwork.

Magnesium bull float: As shown in Figure 2-4, the bottom side of a bull float is slightly bowed in the middle, but is smooth so that cream in the concrete can be brought up and smoothed. This tool is used after striking off the concrete.

Magnesium hand float: The hand float, Figure 2-5, is used in the finishing process to bring cream to the surface. The tool can also put stones into the concrete, prepare the surface for edging,

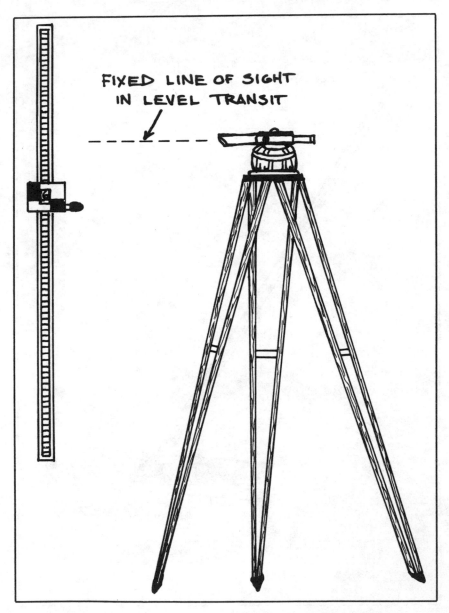

Level transit
Figure 2-2

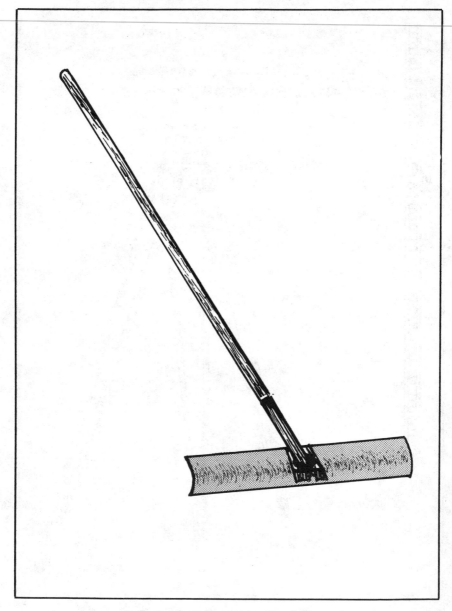

Long-handle concrete placer
Figure 2-3

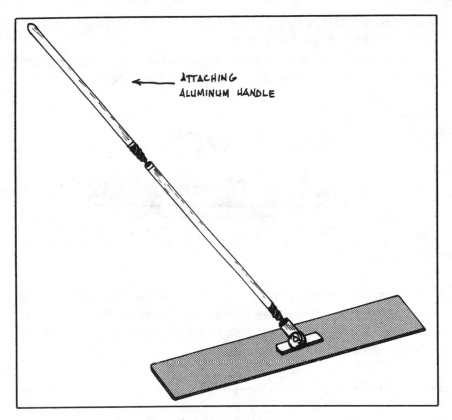

ATTACHING
ALUMINUM HANDLE

Bull float
Figure 2-4

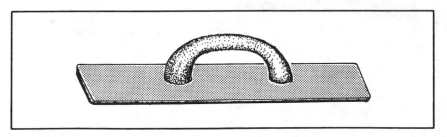

Magnesium hand float
Figure 2-5

13

and be used with the steel trowel to prepare the surface for a final finish if hard troweled or broomed.

Trowel: The trowel, shown in Figure 2-6, is used in the final step of the finishing process to produce a hard, smooth finish.

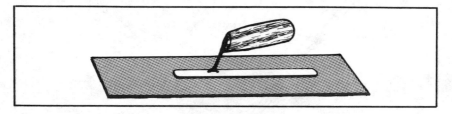

Hand trowel
Figure 2-6

Cement finishing broom: Shown in Figure 2-7, this broom is used in the cement finishing process to put a lined finish on the concrete. A bull float handle screws into an attachment on the broom, and when the concrete becomes tacky, the broom is pulled

CURB BROOM

THIS PIECE BOLTS
TO TOOL FOR
ACCEPTANCE OF
HANDLES AS USED
IN FLATWORK

Cement finishing broom
Figure 2-7

across the surface of the concrete. Brooming gives a finish that provides good footing on exterior cement. The best broom is one made of camel hair, because it puts a more consistent and even finish on the concrete. The broom may also be used without handles to broom curbs and gutters.

Outside step tool: This custom-made tool, shown in Figure 2-8, is used to edge the top and smooth the face of steps. Its flat stock is ⅛ inch thick and 16 inches by 14 inches with a ¼ inch to ¾ inch radius bent at its axis.

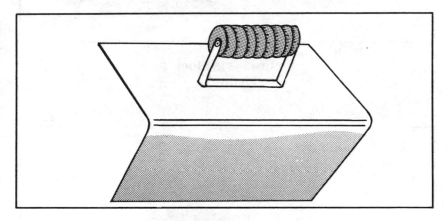

Outside step tool
Figure 2-8

Inside step tool: (See Figure 2-9.) The inside step tool puts a smooth radius on the part of the step where the top and the face meet.

Jointer: The jointer, shown in Figure 2-10, is used for making crack-control joints in concrete. A jointer grooves the surface of the concrete approximately one-half to three-quarters of an inch in depth. This tool also can be used to place designs or patterns on a surface.

Jitterbug: (See Figure 2-11.) The purpose of the jitterbug is

15

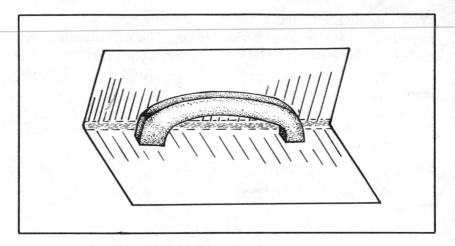

Inside step tool
Figure 2-9

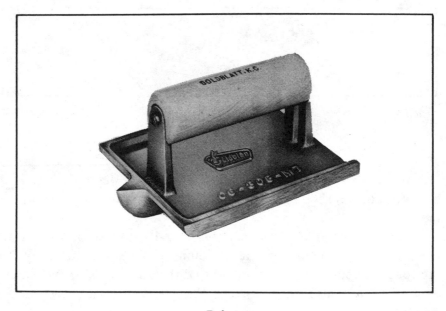

Jointer
Figure 2-10

Jitterbug
Figure 2-11

threefold: it brings cream to the surface of concrete, ties in loads of concrete placed at different times, and flattens the surface of the concrete when pliable.

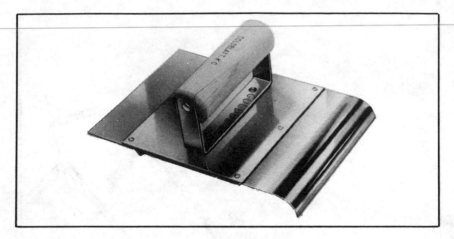

Edger
Figure 2-12

Edger: An edger is used in the finishing process to make a small radius at the outside edge of the concrete, as shown in Figure 2-12. This rounded edge improves the appearance of the concrete and also makes the edges last longer.

Cement edger and safety step groover: Like the outside step tool, the device shown in Figure 2-13 puts an edge on the top of the step. In addition, it grooves the top of the step to provide better footing.

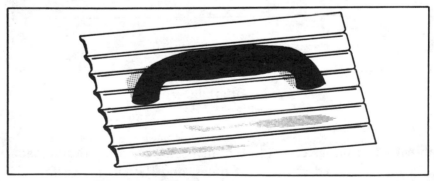

Cement edger and step safety groover
Figure 2-13

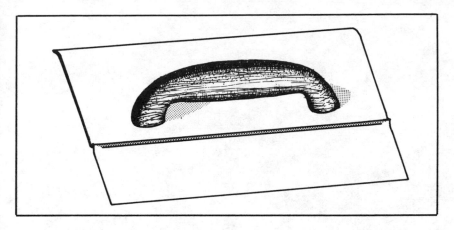

Curb and sidewalk tool
Figure 2-14

Curb and sidewalk tool: As shown in Figure 2-14, this tool forms and edges curb. It also puts a control groove in to separate the sidewalk from the curb if they are at the same elevation.

3
Curbs and Gutters

Many contractors turn down curb and gutter work, considering it a specialty they are not qualified to handle. Though there are contractors who specialize in large curb and gutter jobs, any experienced general contractor can and should be able to place a short run of curb or gutter. Such ability can lead to other projects and can increase profitability. For the residential concrete contractor the ability to do curb and gutter work can lead to more foundation and flat work. And the commercial concrete contractor who can do curb work should experience little difficulty in finding several good-sized curb jobs every year. Thus, curb work is a specialty not to be ignored. It can be good business for many contractors, and understanding the basic techniques isn't difficult, even if you've never laid out and poured a curb before.

Curb Types
To perform curb and gutter work it is essential to know the basic

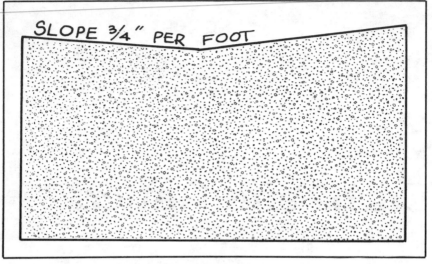

Depressed curb
Figure 3-1

curb and gutter types and their standard design uses. These curbs are the depressed curb, the barrier curb, and the mountable curb.

The depressed curb is one which does not have a high back (curb portion). It has a gradually sloping surface, making it suitable for approaches and driveways, or when a lower back elevation is required. (See Figure 3-1.)

The barrier curb is a curb and gutter combination with a highly restrictive design. The back edge of the curb is perpendicular to the roadway to limit vehicular traffic. These curbs are used in both commercial and residential areas. (See Figure 3-2).

The mountable curb is much like a barrier curb except that it has a rounded back portion. (See Figure 3-3.) Mountable curbs are designed to allow vehicular entrance and are used primarily in subdivisions and parking lots.

These curb and gutter types can be designed to accommodate various combinations of width, slope, thickness and height, depending on local requirements and regulations.

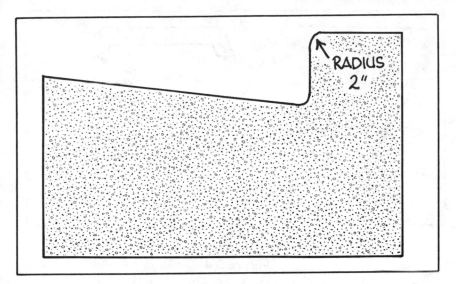

Barrier curb
Figure 3-2

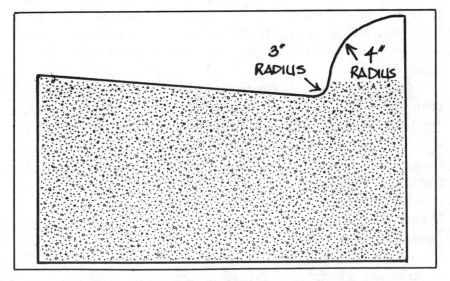

Mountable curb
Figure 3-3

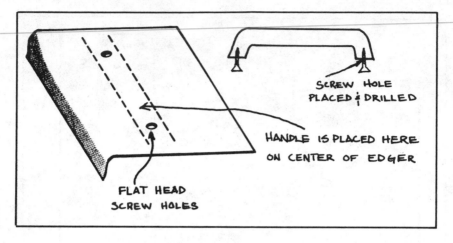

SCREW HOLE
PLACED & DRILLED

HANDLE IS PLACED HERE
ON CENTER OF EDGER

FLAT HEAD
SCREW HOLES

Curb edger
Figure 3-4

Tools

As in any construction work, tools play an important part in the placement of curbs and gutters. The tools discussed below are commonly used in most curb and gutter work. As many of these tools must be custom made, this chapter will also explain how to make them.

Curb Edger This is a custom-made tool used to edge the flag portion of the curb. (See Figure 3-4.) It is made from flat stock sheet metal 1/8 inch thick. Recommended dimensions are 10" x 6" or 7" because these best accommodate different curb applications. The tool is made by bending one end of the sheet metal to form approximately a ½-inch radius. Two holes are then drilled through the metal to accept flathead screws. A handle can be purchased at a local supply or hardware store and attached to the center of the top of the tool.

Curb Top Edger This is another custom-made tool, molded by hand with a hammer. It is used to edge the top of the curb. A form of wood is cut and contoured to the necessary radius. These

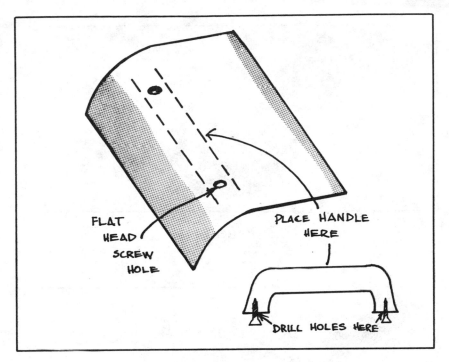

Curb top edger
Figure 3-5

dimensions can be obtained from the curb specifications. A flat stock of sheet metal 1/8 inch thick, 10 inches long and 6 inches wide is moulded to the contoured wood form. A handle can then be attached to the top of the tool. (See Figure 3-5.)

Butterfly The butterfly is a custom-made tool used to complete the forming of the flow line. (See Figure 3-6.) It is best made by first constructing a wood form of the curb and then hammering flat stock sheet metal (1/8 inch thick, 14 to 16 inches long and 12 inches wide) into that shape. This will produce the configuration necessary to form the flag, flow line and face of the curb. Place a handle at the radius bend centered at the curve. This will allow pressure to be exerted evenly over the surface of the curb.

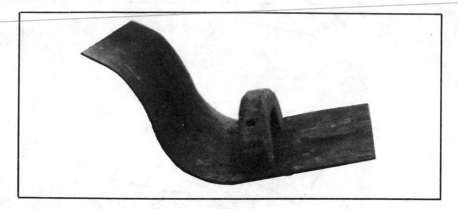

Butterfly
Figure 3-6

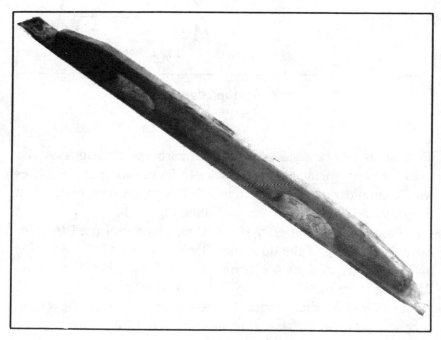

Straight edge with level
Figure 3-7

Straight edge with Level This tool is used to form radius curbing on straight curb runs. The straight edge is made from standard 2" x 4" stock, preferably hardwood. Cut the stock to a 38-inch length and form two hand holds about 12 to 14 inches from each edge as shown in Figure 3-7. The level is made from 1" x 2" or 1" x 3" stock cut to a 48-inch length. This piece is attached to the straight edge, using about 5 flat screws.

Mall A 10-pound mall is usually used for curb framing. The lighter weight makes framing easier and less tiring. The handle is also shorter to make pounding stakes easier.

Expansion Joint Expansion joints are used to control cracking. The specifications will determine the intervals at which the joints are to be placed. These joints are made of a fibrous material and can either be purchased or home-made. Special connectors can be purchased if smaller joints are used. Make sure the expansion joints you use conform exactly to the configuration of the curb. (See Figure 3-8.)

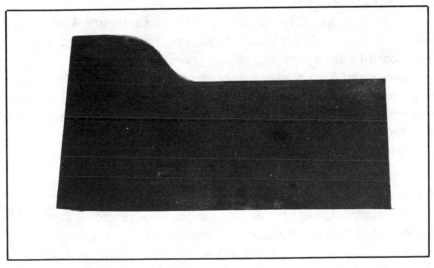

Expansion joint
Figure 3-8

Mountable curb gauge
Figure 3-9

Curb Gauge This tool is used to set the front boards of the curb frame. It determines the exact width and height of the curb and assures that the curb will be level. Figure 3-9 shows a mountable curb gauge.

Plate Tamper This tamper, as shown in Figure 3-10, has 14-inch wide plate and is placed inside the framed backboard and frontboard to allow tamping of the granular stone base.

Rebar and Sleeves Many states require that rebar and sleeves be placed in curb expansion joints. The rebar fits into the sleeve which must be greased to facilitate movement. (See Figure 3-11.) Two holes are placed in the expansion joint, so when the concrete is poured the rebar will be held straight in the joint.

Other materials are also frequently used in curb and gutter placement. Burlap can be used to protect curbs from rain and freezing temperatures. If the weather is very severe, a layer of straw may be placed on top of the burlap for added protection.

12 penny nails are used for framing, and 8 penny nails are used to place radius boards.

Plate tamper
Figure 3-10

SETUP PROCEDURES

Curb work seldom includes any significant grading. The excavating and grading will usually be done by a general paving contractor before the curb crew comes on the site. Still, it's good practice to check the grades before any work commences. This ensures that no additional excavation will be required other than filling with a granular base, if necessary.

At the job site, notice the series of survey stakes positioned at specific intervals where the curb is to be placed. These are the grade stakes and have been set by surveyors to identify where the

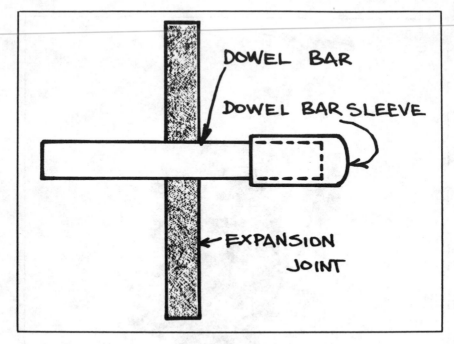

Rebar and sleeve
Figure 3-11

curb will be poured. They are called *hubs* and mark a point on the ground which is also shown on the plans.

These stakes are not placed exactly where the curb will be poured. Instead, they are offset from the curb line by several feet. This allows placement of the curb forms without disturbing the hubs. The offset identifies how far the back of the curb will be from the given grade stake. The *cut* or *fill* is also noted on the grade stake. The cut or fill mark tells how much the level of the top back of the curb is above or below the level of the top of the grade stake. (See Figure 3-12.)

Engineer's Measurements
Notice that all distances are recorded in engineer's scale. If you

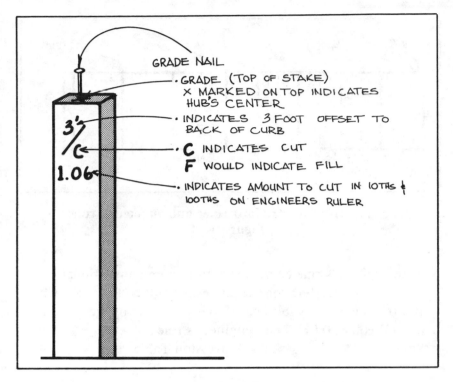

GRADE NAIL

· GRADE (TOP OF STAKE)
X MARKED ON TOP INDICATES
HUB'S CENTER

· INDICATES 3 FOOT OFFSET TO
BACK OF CURB

· **C** INDICATES CUT
F WOULD INDICATE FILL

· INDICATES AMOUNT TO CUT IN 10THS &
100THS ON ENGINEERS RULER

3'/C
1.06

Grade stake
Figure 3-12

haven't used an engineer's rule, all you need to know is that engineering measurements are made in feet, tenths, and hundredths of a foot rather than inches and fractions of an inch. This makes adding and subtracting grades much easier. The following example explains how to read an engineer's scale.

Say a hub is marked **cut .85**. On the engineer's rule this means 8 tenths and 5 hundredths of a foot. A reading of more than one foot might look like this: **2.85**. The 2 means 2 feet and the .85 again indicates 8 tenths and 5 hundredths of a foot. Readings for curb work will always be measured this way. It is the accepted form of measurement on excavation plans also. Figure 3-13 shows

31

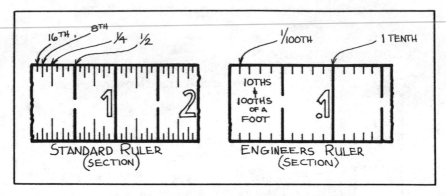

Comparison of standard ruler and engineer's ruler
Figure 3-13

how an engineer's rule compares with a more conventional ruler.

To convert readings on the engineer's rule to feet and inches, follow the conversion table below. When converting, keep in mind that 1/8" equals 0.01' on an engineer's rule.

Sample Conversion Table

1"	=	0.08'
2"	=	0.17'
3"	=	0.25'
4"	=	0.33'
5"	=	0.42'
6"	=	0.50'
7"	=	0.58'
8"	=	0.66'
9"	=	0.75'
10"	=	0.83'
11"	=	0.92'

Committing this chart to memory can make thinking in terms of tenths and inches second nature. But the easiest way to perform these measurements is to purchase an engineer's rule of your own.

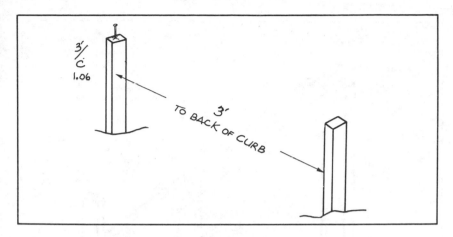

Setting a stringline
Figure 3-14

Setting Stringline

To begin, look at the first grade stake and read the offset. Put a nail in the top center of the grade stake and measure at a right angle from the line of stakes the given offset distance. This point is the exact back of the curb. Position a stake on the inside of this point. (See Figure 3-14.) This stake is the first support for the stringline that will mark the back line of the curb.

Next, set the height of the top of the curb at that point. (See Figure 3-15.) Use a 6-foot level to transfer the height at the top of the hub to the stringline stake you have just placed. Mark this level on the stringline stake as shown. Then note the amount of cut or fill marked on the grade stake. Place the engineer's rule on the stringline stake and measure up or down from the mark the distance indicated for cut or fill. Measure down for cut and up from the mark for fill.

Transferring the curb level to the stringline stake is easier if you drive a nail in the top of the hub. Over-driving will disturb the hub, so drive the nail gently. Then place the end of the 6-foot level against the nail. This helps hold the level at the center of the stake.

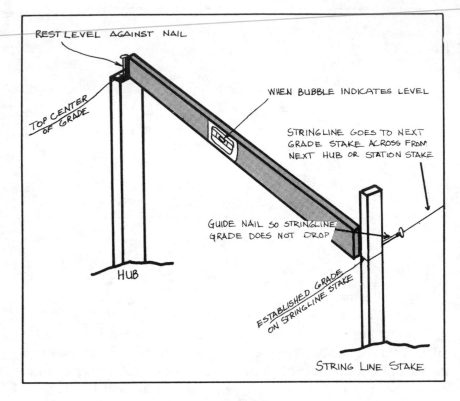

REST LEVEL AGAINST NAIL

WHEN BUBBLE INDICATES LEVEL

TOP CENTER OF GRADE

STRINGLINE GOES TO NEXT GRADE STAKE ACROSS FROM NEXT HUB OR STATION STAKE

GUIDE NAIL SO STRINGLINE GRADE DOES NOT DROP

HUB

ESTABLISHED GRADE ON STRINGLINE STAKE

STRING LINE STAKE

Transferring height at the top of the hub to the stringline stake
Figure 3-15

Some surveyors put an "X" in the top of the grade stake to mark the exact center, as shown in Figure 3-14.

Once the top back of curb has been marked, drive a nail at this point on the stringline stake. Continue setting stringline stakes and marking the top back of the curb for as much work as you plan to do for that day. Then connect all the stakes with stringline stretched from one nail to the next. The nail keeps the string from dropping from the grade mark. Once this has been completed and checked at each station, the forming lumber can be laid out.

LAYOUT, FORMING AND POURING

An organized layout of both stakes and forms is essential to handle curb work efficiently. An experienced curb crew can be very efficient, and with a little planning and the right equipment, hours of work can be saved. Ideally, the layout should be handled in the following way.

A professional curb crew will use a truck with a 10- to 14-foot dump body or a one ton dump truck. The truck is used to pull a 16-foot bed trailer. The trailer carries the framing lumber and the truck carries the stakes.

Usually 36-inch stakes are used to support the forms. Shorter stakes could be used, but longer stakes are preferable. Most curb is placed on clay or firm soil. Pounding the stakes weakens the tops so they must be trimmed occasionally. A longer stake means longer stake life. Also, the backboard (the form at the back of the curb) requires a taller stake, because the forms are higher at the back than at the front. Braces for the backboard stakes have to be longer yet. They must be set at a 45 degree angle to provide support for the line stakes that hold up the backboard. Use any shorter stakes you have for the frontboard.

Only good quality stakes should be used. They are more durable and will reduce your material cost. In some areas metal or even plastic stakes have replaced wood for this type of work.

Lumber for the forms may be 2 x 8's, 2 x 10's, 2 x 12's, 2 x 14's, or even 2 x 16's, depending on the curb height. 2 x 16's are most frequently used. This lumber is expensive, but if used properly can last for many years. One method of promoting longevity is *bolting* which reinforces and stabilizes the board. (See Figure 3-16.)

Bolting requires holes drilled through the lumber width and spaced at set intervals. Six 3/8-inch stove bolts should be fitted into the lumber as shown in Figure 3-16. Keep the bolt holes back about a foot from the end of the board so the ends can be trimmed when they begin to split. Place a washer and nut on the bottom

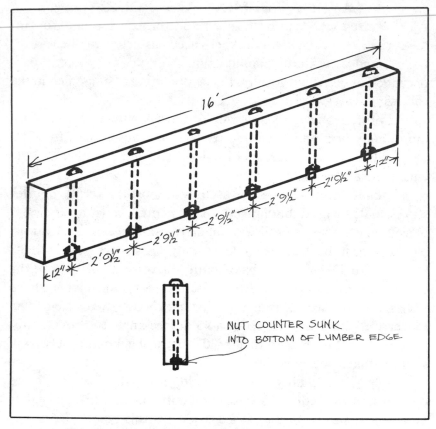

16′

12″ * 2′9½″ * 2′9½″ * 2′9½″ * 2′9½″ * 2′9½″ * 12″

NUT COUNTER SUNK
INTO BOTTOM OF LUMBER EDGE

Bolting to reinforce and stabilize form lumber
Figure 3-16

edge of each board and tighten securely. The nut should be countersunk about 1/4 inch below the bottom surface of the board and tightened enough to put stress on the lumber.

Bolting keeps the lumber from splitting even after years of nailing and hard use. Another advantage is that the lumber becomes more rigid and holds its shape better under the weight of the concrete. These two points are very important in keeping form lumber and labor costs to a minimum.

Laying Out Materials

Load the lumber onto the truck. Make sure that the backboard lumber is placed on the right side of the truck bed so it can be pulled off easily. The usual method is for one man to drop off the required amount of stakes — sixteen for each 16-foot board — while another man pulls off the backboard lumber and lays it end to end inside the stringline. Other members of the crew can begin framing the backboard and placing line and kicker (brace) stakes right behind the men who are unloading and setting up the lumber and stakes.

Once the backboard lumber is laid out, the men in the truck and trailer return to the beginning of the curb line and start unloading the frontboard framing material. Use 24-inch stakes for framing the frontboard. As before, the men unloading the form lumber drive the line stakes, frame the curb front and pound the 24-inch kicker stakes.

A good crew size for forming the curb front and back is four men: a lead man who sets the forms, a stake pounder, someone to place the kicker or support stakes and someone to nail stakes and form lumber together.

Setting Mountable Curb Forms

The lead man starts from the stringline that has already been set. This line marks the top back of the curb. He sets the backboard stakes so the form lumber will be flush against and level with the stringline. Then he places the frontboard at the proper elevation and spacing. A curb gauge is essential for this work. The gauge helps measure the proper width of the curb and also establishes the exact elevation of the frontboard. (See Figure 3-17.)

The curb gauge is held with the level end flush against the backboard and the level resting on the backboard. The notched end is raised until the gauge reads level. This is the correct position for the frontboard. A different curb gauge is used for every different type of curb.

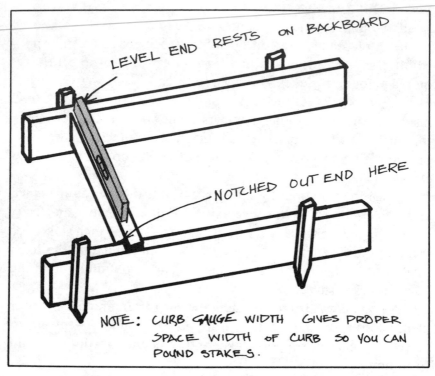

LEVEL END RESTS ON BACKBOARD

NOTCHED OUT END HERE

NOTE: CURB GAUGE WIDTH GIVES PROPER
SPACE WIDTH OF CURB SO YOU CAN
POUND STAKES.

Determining front board height and proper curb width
Figure 3-17

The man pounding stakes also needs a gauge to check spacing as the stakes are driven. (See Figure 3-18.) The nails at the left end of this gauge rest on the backboard. When the gauge is held snug between the backboard and the frontboard, the frontboard is in the right position. A stake is then placed against the outside of the frontboard. Drive three line stakes for each sixteen foot length of form lumber.

When three stakes are in place, the lead man and stake pounder move on to the next length of form to repeat the process. When the next form board is in place set the joint stake where the two lengths of form join. (See Figure 3-19.) Nail joint stakes from

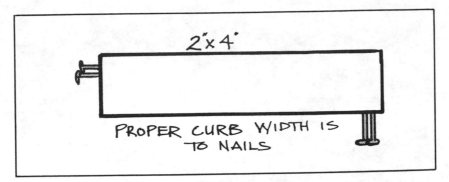

Curb width gauge
Figure 3-18

the inside of the form where the concrete will be poured. All other stakes are nailed from the outside of the form.

The next step is to place the kicker stakes which brace the line stakes. (See Figure 3-20.) The kicker stakes are driven at a 45

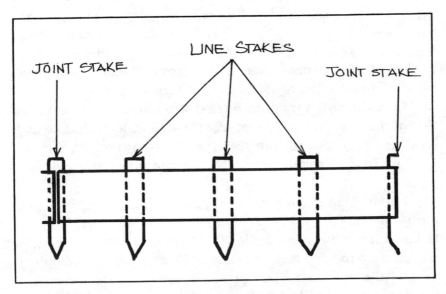

Placement of line stakes and joint stakes
Figure 3-19

Kicker stakes brace the line stakes
Figure 3-20

degree angle. The last man on the crew nails all the framing boards to the line stakes and checks the kicker stakes to make sure they have been placed properly. He too needs a spacing tool to make a final check of alignment. Coated nails are best because they allow some adjustment and tightening of the kicker stake.

Once the curb is framed, the next step will usually be application of the stone base. The type of base used varies with state and local requirements and with the type of base material available. Once the base is in place, it is ready to be tamped with a plate compactor.

Forming radius curb and gutter takes a little more time. The most common form material for radius curb is Masonite, but steel radius forms are also available. The technique is the same as in straight curb framing, but more line and kicker stakes are required to hold the true curve. This makes it easier to strip the boards when the job is finished. Figure 3-21 shows a completed radius curb form.

Forming radius curb and gutter
Figure 3-21

Forming Barrier Curb

A barrier-type curb with a vertical face requires an additional step. Placement of the backboard and frontboard is the same, but the inside of the curb face must be formed with a faceboard. The faceboard form can be a 1-inch board or a steel form and is usually sixteen feet long. The width of the faceboard will be the same as the width of the curb. For example, in a B-6-12 curb (barrier type with a 6-inch high face and a 12-inch height at the curb back), the faceboard must be six inches wide.

When the backboard and frontboard are in place, the faceboards are placed at the same elevation as the top of the backboard. Place a metal spacer between the backboard and the

inside of the faceboard and secure it with an adjustable curb
clamp. (See Figure 3-22.) The spacer holds the faceboard in place
against the clamp until the concrete is poured. Use either 18- or
24-inch adjustable bar clamps. Be sure to oil the clamps before and
after each pour, since contact with the concrete will ruin them very
quickly.

A spacer and clamp are usually needed at approximately four
foot intervals. Make sure that the dimensions of the spacer conform to the barrier curb specifications. Place a clamp and spacer
six inches in from the end of each faceboard and make sure they
are evenly spaced. Figure 3-21 shows the clamps in place. One
spacer is required at every clamp.

Adjustable curb clamp
Figure 3-22

Pouring the Curb

Pouring the concrete is a critical part of the job. An experienced
and well-organized crew can produce good results with little
wasted motion.

Before pouring, spray both the backboards and the front-

Condensing concrete as it's poured
Figure 3-23

boards with a commercial concrete form oil. This keeps the cement from sticking to the boards and makes stripping easier.

When pouring curb by hand, the proper slump is 3 to 3½ inches. Your concrete supplier should be able to supply the right mix for your jobs. You may prefer a slightly more or less workable mix, depending on the curb design. Generally, the dryer the concrete mix, the easier it is to finish.

Don't worry about the concrete setting up too fast. With a curb crew of six or seven men, finishing each section takes very little time. Often, the curb can be finished as quickly as the concrete truck can complete the pouring.

Usually concrete is poured directly from the chute and condensed against the backboard and the frontboard with a shovel. (See Figure 3-23.) This keeps the face and backboard concrete from *honeycombing*. Honeycombing occurs when the gravel in the concrete is not mixed well with the creamy material in the mix. The gravel remains visible on the surface of the finished curb. This is a common problem if condensing is not done properly, and patching will be required to finish the job.

Place the shovel in the concrete next to the form boards. The back of the shovel should be against the form. Move it up and down so that cream from the concrete can seep between the shovel and the form.

FINISHING

As this is being done, a member of the crew shapes the curb with a flat shovel, so a straightedge can be worked to form a rough version of the curb design. The man using this straightedge must work carefully. He is forming the shape of the finished curb. Consistency is important. Work the straightedge back and forth with two hands to level and shape the concrete and bring the cream to the top. The tool is pushed and pulled the same way as a hand float would be used on a concrete slab.

Condensed, rough graded, straightedged and shaped, the curb is now ready to be finished. An edger is passed along the *flag* or street edge of the curb. It cuts a nicely rounded corner into the concrete at the point where it meets the form. This prevents the concrete from chipping once it has dried. Make sure that the tool is slanted toward the flow line, even if you are forming a reverse gutter curb which drains toward the street.

At the same time, run the edger along the outside edge of the curb. This edger is run flat rather than slanted. Figure 3-24 shows how a mountable curb looks with both edges shaped. It is important to follow the board shape closely and run the edger as consistently as possible.

At times the concrete on both edges may become difficult to work. Adding a slight amount of water should restore workability.

Mountable curb with both edges shaped
Figure 3-24

Use a plastic sprinkling can, but be careful not to let water collect in the flow line of the gutter. This is where the third finisher must work with a butterfly.

As custom-made tools go, the butterfly is one of the most interesting. With this tool, the flow line is formed to precision and the curb becomes smooth and ready for brooming. Figure 3-25 shows the butterfly in the center and two other trowels used to form a mountable curb.

The motion of the butterfly is the same as that of a basic edger. When running the butterfly, watch the space between the backboard and the edge of the tool. Hold that space consistent while running the tool on the curb. Once the butterfly is run, the curb is ready for brooming.

A natural-hair brush is used for brooming the curb. The brush should be dipped in water and kept damp at all times. For best results, use a continuous wiping motion starting at the street edge of the curb. Always broom the curb at a 45 degree angle. (See Figure 3-26.) At rough spots, lightly sprinkle the curb with water and then brush. If the curb is a little too wet, keep the broom drier.

45

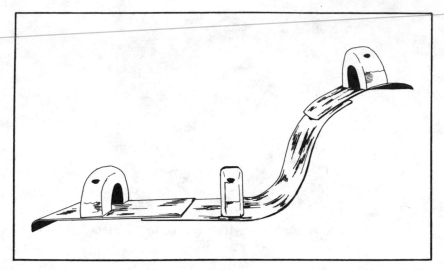

Tools to form a mountable curb
Figure 3-25

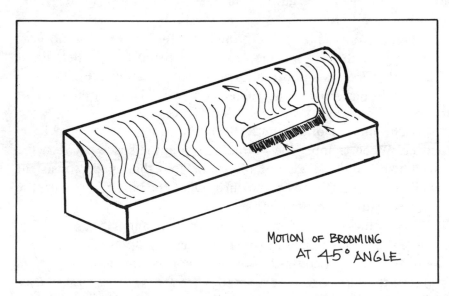

MOTION OF BROOMING
AT 45° ANGLE

Broom curb at 45 degree angle
Figure 3-26

Good brushing means the difference between a nice looking curb and a crude job, so use a continuous wiping motion with each stroke.

Barrier Curbs and Gutters

The concrete placement for barrier curbs requires different tools and procedures. Concrete pouring is the same as for a mountable curb. But as the concrete is poured, each clamp is loosened slightly one-eighth of a turn. The metal spacer is removed, allowing the concrete to fill the faceboard and upper half of the backboard. The spacers were used only to hold the faceboard away from the backboard until the void was filled with concrete.

Once the concrete is poured and the spacer removed, a finisher uses a hand float to shape the curb top and bring up the cream. Then front and back edgers can be worked to form a smooth surface on the curb top. The top curb edger is used up to and between the clamps on both the back and inside curb edge. (See Figure 3-27.) Finish the top edges of the barrier curb as quickly as possible to allow enough time to pull the clamps and faceboards before the concrete sets.

Once the concrete will hold its shape, the clamps and faceboards are no longer needed and can be removed.

Finish the curb with an "L" shaped tool called a 90 degree butterfly. (See Figure 3-28.) Use this butterfly with the same inside and downward pressure as used with the mountable curb butterfly. It may be necessary to bring up the cream with a straightedge at the flow line point if any stones are present or if the butterfly does not do an efficient job.

The curb is now ready for brooming. Start at the street-side edge of the curb and stroke at a 45 degree angle over the curb, finishing at the top curb edge.

Edging the top of the curb
Figure 3-27

90 degree butterfly
Figure 3-28

Form Stripping

A crew should be able to strip and load the lumber and stakes as quickly as they were able to frame and pour the curb. With a crew of six men, one trailer, and a dump truck or stake bed truck, the stripping of forms can proceed quickly and efficiently.

The first two men strip the kicker and line stakes, putting them in small piles in front of the curb as they go. A good, strong all-metal hammer is ideal for knocking out the stakes. Next, two other men strip the front and backboards, pull nails in these boards, and place them just in front of the finished curb. Take care to avoid hitting the wet concrete with the form material.

The last two men on the crew are used for cleanup. It is their job to take the nails out of the stakes and load the form lumber as quickly as possible. One man drives the truck and trailer and loads the stakes in the dump body. The other man loads the form lumber, making sure that the backboard lumber is put on the trailer nearest the driver's side. The frontboard lumber goes on the passenger side of the trailer.

By this time the lead stripper should be finished and can fall back to help the others load. With practice, the crew can become very efficient at this stripping procedure in a short period of time.

Avoiding Weather Problems

As with all outdoor concrete work, weather can be a problem. If rain is expected, cover the freshly poured concrete with burlap. The material used should be at least three feet wide and 150 feet long. If rain begins, cover the finished curb immediately and stop pouring new curb. The burlap will protect the finish of the already-poured curbing.

Some specifications require that a sealer be put down after the curb is placed. Usually this sealer will be tinted with red dye for inspection purposes. If sealer is required, it can be applied with a common sprayer after the curb has been brushed.

CURB MACHINES

The curb machine is a new development designed to aid and simplify curb and gutter placement. At present the machines are costly and cannot be used in all applications. Some still require technological refinement, and special crews must be assembled to set up and run the equipment. But they do offer certain advantages and are worth considering if you plan to do a lot of curb work.

Curb machines are self-propelled units that use a ram or vibrators to form the curb and move the machine forward as more concrete is poured into it. The unit follows a static line or stringline which has been set with a metal pin to hold the line on grade. (See Figure 3-29.)

The curb is shaped and formed perfectly without the use of forms or stakes: a process known as slipforming. Slipforming eliminates the need for setup and stripping since forms are not used, and provides a more uniform and consistent curb than conventional methods. Finishing methods for most machines, however, remain the same; the curb must still be smoothed with a butterfly, edged and brushed. (See Figure 3-30.)

Curb machine
Figure 3-29

Slipformed curbs must still be finished by hand
Figure 3-30

4
Driveways and Sidewalks

For the residential concrete contractor, driveways and sidewalks constitute the major part of exterior concrete flatwork.

To set up a driveway, determine four elevation points. Most of the time, these points are already given. The garage floor elevations are a starting point. Remember to keep the driveway elevation ½ inch lower than the garage floor elevation by holding concrete ½ inch below the finished floor grade. The setup for a rectangular or slight straight-angle driveway is shown in Figure 4-1. The procedure is basically like that of a patio, with stringlining around four stakes. Pound the first two stakes 16 feet apart or a garage door width apart if other than 16 feet. Assume a straight line on one side and pound a third stake next to the sidewalk. Then measure 16 feet from the stake and pound another stake into the ground next to the sidewalk. Once these four points are established, set up the stringline flush with the sidewalk ½ inch below the

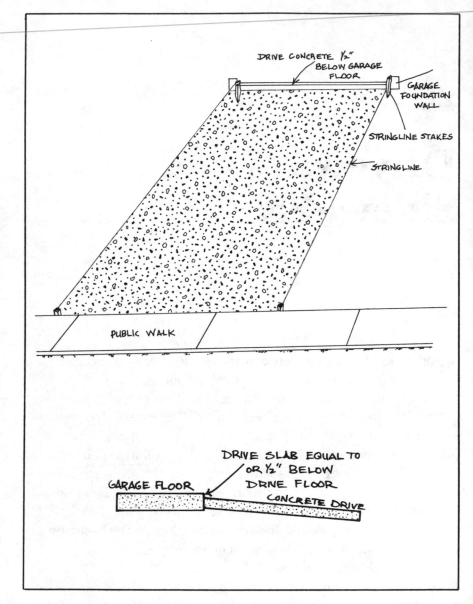

Setup for a rectangular or slight straight angle driveway
Figure 4-1

garage floor. At this point, make sure the drive is on the property by checking for the property pins. Then dig out the driveway 6 inches to 1 foot deep (depending upon how much stone base is required by the specs). Set up and form framing lumber and place stone to the bottom of the framed lumber.

For curved driveways, use 2 x 4's for the straight sections and masonite to form the curves. Connect the masonite to each 2 x 4 with eight-penny nails. Maintain the elevation pitch from the 2 x 4's already framed by connecting a stringline from the top of one 2 x 4 to another. In this way the proper stringline elevation between the framed 2 x 4's can be transferred to the proposed curve shape that is formed with masonite. This helps maintain constant slope so the finished driveway drains well.

If a driveway has minimum fall, pitch the width of the driveway ¼ inch per foot and away from the house. One side of a 16-foot driveway will be 4 inches higher than the opposite side pitched at ¼ inch per foot. If the driveway is a double drive (20 feet or wider), pitch the driveway to both sides with the center as the high point. In this case, frame a screed down the center. When one side is poured, graded and bullfloated, do the other side the same way and then pull out the screed boards. The concrete is now properly pitched at ¼ inch per foot with the center as the high point, as shown in Figure 4-2. You may question how this can be done from a flat garage floor. The answer is to hold the elevation of the driveway ½ inch below the garage floor and keep the drive pitch from the center ⅛ inch per foot, working into ¼ inch within 3 feet of the garage floor.

In some cases, the driveway must be pitched toward the garage floor. The following are two ways to do this correctly.

First, if there is a great amount of pitch back to the garage floor, a drain system will probably have to be installed just in front of the approach. Use a ready-made system with a concrete trough installed with separate grating running the width of the drive.

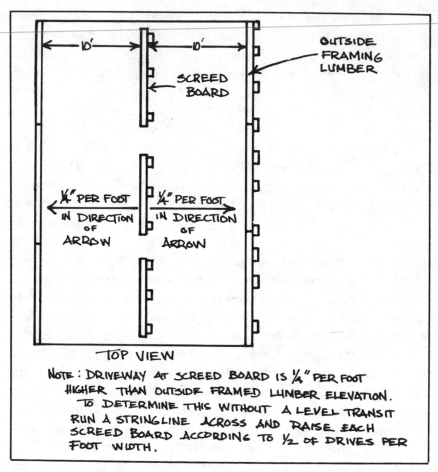

Proper pitch for minimum full driveway
Figure 4-2

From the ready-made system, place PVC or drain tile tubing underground and run to a French drain, culvert or storm sewer. (See Figure 4-3.) Pitch the driveway trough the same as the driveway flatwork so that water will run out to the drainage system setup.

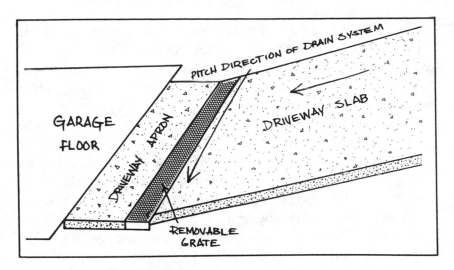

Drainage system when pitch is back to garage floor
Figure 4-3

A less-expensive method of draining water is to install a circular drain with 4-inch flexible hoses that pitch to the selected area. Direct driveway water to this point by creating a cone around the lowest point of the drain. Make sure that the drain cover can be removed from the surface of the driveway to allow removal of mud, twigs or leaves which can clog a system and cause water to back up. To free the drain cover, tap it with a hammer. It will loosen if the concrete has been poured flush to it.

When framing drives with the above drainage systems, set up the driveways with both elevations equal on the outside framed boards so water is directed to the system, or, in the case of the trough system, pitch the drive away from the house at the same pitch as the system to shed water along its course. Remember that where two hard surfaces meet, such as floor to drive and drive to sidewalk, place ½'' x 4'' expansion joints between the two hard surfaces to absorb the expansion.

Sidewalks are among the most interesting of residential work because they can be designed to fit the needs of each individual as well as his home. A front service walk is usually 3 feet in width, side service walks are 2½ feet in width and public walks are 4 to 5 feet in width. These variances are to accommodate differences in use: less-used walks are narrower than heavily traveled public walkways. Specs concerning public walks are always governed by municipal building codes which will be discussed later. A sidewalk is always pitched ⅛ inch to ¼ inch per foot of width. A 3-foot-wide sidewalk is pitched ¾ inch away from the house, a 5-foot-wide sidewalk is pitched 1¼ inches. The exception to this rule is when the fall of the walk is great enough to allow sufficient runoff.

A front service walk always joins a front stoop or doorsill, so it must be held down 5 to 6 inches. This automatically determines the first two starting points for the walk. A front service walk ends usually at a driveway; these are the two end-point elevations. Two other elevation points for the drive are important in the case of the sidewalk making a turn to the drive at a 45 degree or 90 degree angle. These points must be adjusted both to conform to the grade of the ground and to allow a proper pitch and uniform fall. The pitch of the sidewalk is essential to eliminate dangerous situations and serious injuries if water should freeze on the sidewalk.

To set up the walk, place four inside stakes in the ground, with the outside stakes set apart the width of the concrete. Mark the elevations that conform to both the elevation down from the stoop (5 to 6 inches) and the driveway elevation. Set up two more stakes at the pivot point of the sidewalk. Adjust them so that the sidewalk will have a uniform width going both ways. The stake closest to the house will be the only stake that is on the outside when the stringline is set up. The rest of the stringline stakes are on the inside. When setting up the stringline, pitch it ¾ inch lower and away from the pivot point of the house. Frame and nail the 2 x 4 forms without interfering with the stringline stakes. The only stake that will be on the outside will be the inner pivot point corner stake. Frame the 2 x 4's to incorporate this into the walk so that there will not be a loss of elevation. The stringline allows proper pitch to be maintained and the sidewalk to be framed straight and true. In most cases, start the sidewalk 10 to 12 feet from the front of the garage. This allows a car to unload passengers directly onto the sidewalk.

Side service walks can involve more work, but are fairly easy to do if the following basics are kept in mind. A side service walk is usually placed 2 feet from the foundation wall. Two feet is the usual extension of a brick fireplace and allows the sidewalk to be framed flush to the brick. Side service walks that meet a stoop should be held down from the face of the stoop 6 or 7 inches maximum with an expansion joint in between. First measure out from the foundation two feet, put the inside stake at the 2-foot mark (usually the corner of the house on one side) and pound it into the ground. Measure 2 feet 6 inches from this stake, or the desired sidewalk width, put another stake inside and pound it into the ground. At this point, place all stakes inside except at the two inner corners of the walk. If framing square angle corners, leave these two inner corner stakes in and frame with two 2 x 4's.

To make the curve of a sidewalk corner, use a piece of masonite approximately 3 to 3½ feet in length and insert it at the corner. Adjust the masonite to the desired outside radius and nail it from the inside of the 2 x 4 corner with an eight-penny nail. Nailing from the inside eliminates nails in the concrete after stripping. If a totally curved corner both inside and out is desired, frame 2 x 4's and stop 2 feet from the inside corner each way, then connect with masonite. If framing a sidewalk that connects to a patio, keep the elevation equal to that of the patio. Construct the sidewalks flush with the patio and place an expansion joint between them.

If the straight span of walk is very long, eliminate the slight sagging of the stringline by placing two inside stakes, marking elevations, and holding up the string with nails. Mark the elevations on the outside of the stakes. Set up the elevation at the stoop on the inside stringline (6 or 7 inches down from the top). If the ground grade is fairly level, assume that each end has the same elevation. This can be done by using a level transit, leveling over using a 2 x 4 and four foot level, or by using a stringline with a line level attached to it.

Set the outside stringline ¼ inch per foot lower all the way around. Do this by shooting on the outside stakes, leveling over with a 4-foot level, and then by dropping the necessary measurement down.

After all elevations are complete, set up the stringline and dig down 4 inches or the planned concrete thickness, plus the depth of the stone base, if any, that is being put in. If there is going to be 4-inch concrete with a 4-inch stone base, dig below the stringline 8 inches. Lay out the 2 x 4's and stakes and start framing. The reason for laying out the 2 x 4's next to the proposed job is to eliminate unnecessary cutting of good lumber. The boards do not have to fit exactly and they can run past the stringline on one side of the framing.

The process of pouring sidewalks is comparable to any flatwork job. First, pour and grade the concrete. Second, strike off and bullfloat, and third, cut the edges. When laying the control joints, which is the fourth step, make sure to space them the width of the pour. Put in an expansion joint every 10 to 20 feet or as specified. Placement of expansion can be made while the concrete is wet to the touch, or cut 2 x 4's the width of the proposed walk and place the expansion in front of the board. Pour concrete on both sides, pull the 2 x 4 out and grade. Put the joints in with the jointer by placing a 2 x 4 over the width of the work and cut the joint. If cutting to the exact length of the joints, lay the 2 x 4 down after each completed joint and mark the next joint at the end of the 2 x 4 — measuring won't be necessary. Hand float and trowel each square and then broom when concrete is tacky. Always broom concrete perpendicular to the work or across its width. These are points to remember:

1. Always pitch any sidewalk ¼ inch per foot of width unless there is a lot of fall to the walk.

2. Keep the front walk 10 to 12 feet away from the garage where it meets the driveway.

3. The width of the sidewalk determines how far apart the control joints should be placed.

4. A sidewalk should have expansion joints every 15 to 20 feet of walk poured.

5. When framing the sidewalk, always keep in mind the final grade of the ground or top of black dirt.

6. Always give the walk a medium broom finish to assure a good footing during rainy periods.

7. For front service walks on new homes, always dig two postholes (where sidewalk starts from stoop) in the center of the first square. This is a stress point or drop point from backfill.

Customized sidewalks realize a large profit compared to the amount of work involved. If a customized sidewalk is desired, run the edger on both sides when the concrete has a loose tackiness to it to leave a slight ridge on the walk, and then broom in swirls in the center. Another way to customize is to install aggregate patio inserts. Plan the length of the sidewalk and the amount of squares it will have. Purchase one patio block for each square of walk and place it in the center of each square flush in the concrete. Swirl with a brush around the exposed aggregate insert.

Concrete Flatwork Manual

5
Patios

As transportation and fuel costs continue to rise, many people are spending more of their leisure time at home. As a result they are building recreational rooms inside their homes. Outside, the patio provides an area for sunning, relaxing and entertaining. With the proper skills a contractor can produce patios designed to meet the needs of practically everyone. To determine particular preferences, consider questions such as:

- Does the homeowner have a large or a small family?
- Does he like to entertain outdoors?
- Are unusual or custom shapes a consideration?
- Does the homeowner like to plant?
- What kind of exterior landscaping is preferred?

Once the preferences of the customer are known by the contractor, choosing the appropriate patio is easy.

An interesting design that appeals to many people is a basic patio with built-in planters. Frame an open area inside the patio.

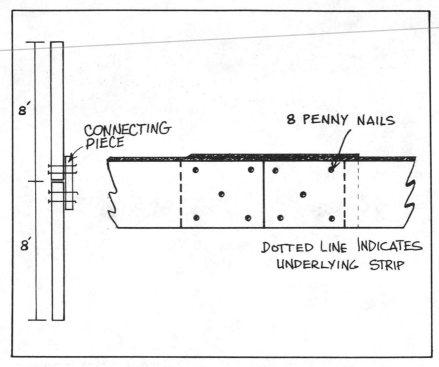

Masonite strip connecting two strips to form a long continuous sheet
Figure 5-1

The concrete will be poured around this space. Frame with 2 x 4's using stakes to hold them from the inside. Set these in accordance with the exact pitch and elevation of the patio to keep everything flush. If circular and semi-circular planters are preferred over straight line-shaped planters, use masonite. Cut a sheet of masonite into strips the thickness of the patio and place end to end in a straight line. Place an 8-inch-long piece of masonite under each joint section where two strips of masonite meet, as shown in Figure 5-1. Drive eight-penny nails into the connecting strips, then turn the long strip over and bend the nails flat. The result is a long continuous flexible strip of masonite.

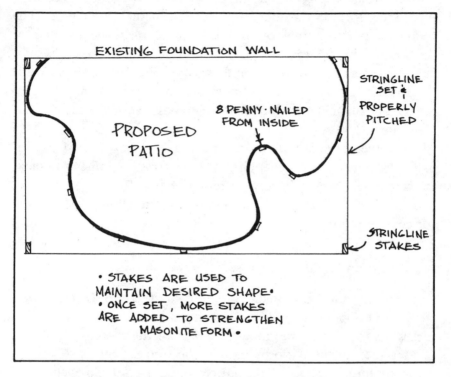

EXISTING FOUNDATION WALL

STRINGLINE SET & PROPERLY PITCHED

8 PENNY·NAILED FROM INSIDE

PROPOSED PATIO

STRINGLINE STAKES

• STAKES ARE USED TO MAINTAIN DESIRED SHAPE.
• ONCE SET, MORE STAKES ARE ADDED TO STRENGTHEN MASONITE FORM.

Placement of stakes for masonite free form patio
Figure 5-2

Dig out the patio in a square and set up the four corner stakes. Take elevations and set up the stringline. Set the masonite shapes inside the square and level over to the stringline so proper pitch is attained, or shoot in with the transit different points of masonite that correspond with the stringline. Pound in the stakes at key points as shown in Figure 5-2. Then raise the masonite to the proper elevation by leveling. Now reinforce the masonite with the other stakes. Use eight-penny nails to nail the masonite to the stakes from the inside out. In most cases, one sheet of masonite (4' x 8') will be enough for any shape, though more intricate designs require more sheets.

Exposed aggregate shapes about 2 inches thick with colored exposed stone on the surface can be used to create custom designs. Place them flush in the concrete before the actual finishing. After concrete is placed, struck off and bullfloated, place the inserts into the concrete by digging out the excess in the shape of the approximate insert size. Adjust up or down and pitch accordingly with the existing cement. Finish in the normal manner. Surface designs can also be made with a jointer in an endless array of design combinations.

Another popular addition that can be incorporated into a patio design is redwood stripping. Set up strips on dollies and tack them to side framing in any pattern selected. Pour the concrete and finish in the usual manner. The strips also eliminate the need for control joints.

A review of the eight steps for setting up a patio is as follows:

1. Keep the patio 6 or 7 inches down from the stoop. If there is no stoop, do not pour the patio on existing siding, but between 1 to 4 inches below, depending upon the grade of the ground.
2. Measure out the patio by placing the four stringline stakes.
3. Set up the stringline for the patio around the outside of the four stakes.
4. Shoot in patio elevations, starting at the stoop of the existing structure. Transfer this starting elevation to the two stringline stakes at the existing structure.
5. Next pitch the patio ⅛ to ¼ inch for each foot of the length of the patio. If the patio is 12 feet, pitch the patio 3 inches lower than the starting elevation at ¼ inch per foot (divide 3 into 12). If a level transit is not available, use a level, or even better, a line level. To use a line level, determine first the starting elevation, then level the stringline all the way around the patio. Drop the stringline 3 inches down (the patio is pitched ¼ inch per foot). Another way to pitch is to use a 2 x 4 the length of the patio. Put the level on top of

it and hold the 2 x 4 to the first designated elevation. Level over to the other stringline stakes. Drop the mark down to whatever the pitch should be.

6. Frame the 2 x 4's to the stringline and put stakes next to the 2 x 4's to hold up the board as well as to keep it aligned with the stringline. Nail with duplex nails from the outside of the stake to the 2 x 4's, keeping the nail head out of the stake.

7. Place the stone in the patio after it is framed and spread it out to the bottom of the 2 x 4's. Make sure the grading is accurate or more concrete will be used than was figured when ordered.

8. Place the wire just before the cement truck arrives. Back bend the wire and overlap and tie as explained earlier. The reason the wire is placed just before the concrete placement is because rusted wire is dangerous. If you are handwheeling the patio, leave one side of the patio open and after the concrete is poured, frame it back up and stake it.

Concrete Flatwork Manual

6

Garage and Basement Floors, Steps and Stoops

Although commercial work is more advantageous to the concrete contractor, residential work is often more available. Most residential concrete work involves garage floors, basements, steps and stoops. Before starting any work, check with the municipal building and zoning departments for any applicable codes for the floors.

To construct a new garage floor, first shoot in the floor allowing ¼" to ⅛" pitch per foot from front to back. Indicate this by marking the proper elevation at the four corners of the foundation wall so that chalk lines can be snapped. Note the exact location of the finished floor on the inside perimeter of the foundation wall. Generally, a garage floor at the back of the foundation is lower in elevation than the finished floor at the back wall. Stretch and snap successive lines from the corners of each side to the front corners to indicate the pitch of the finished floor.

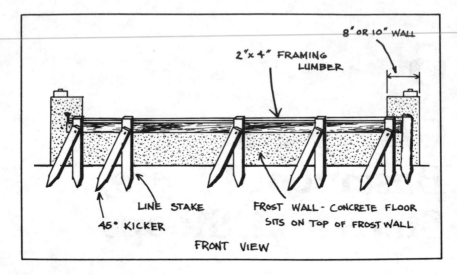

**Completed framing for garage floor slab at door entrance
Figure 6-1**

If the wood door trim for the front of the garage has already been installed, make sure the front of the finished floor is flush against the bottom of this trim. If the garage is roughed in the floor will be approximately 7 feet below the bottom of the header. It is best to the use a level transit to shoot in the corner elevations, but you can use a line level if necessary.

Once the snapped chalk lines are marked, begin digging out unnecessary dirt or clay so that a stone base can be put in. Before you place any stone, flood the dirt floor by digging a hole in each corner in the clay and running water continuously by hose into the holes at a slow rate. By doing this, any loose ground is compacted to produce a quality longer-lasting floor. Place a crushed limestone base to the required depth throughout the floor area while still allowing for the thickness of the concrete floor.

At the front of the garage, run a stringline to two stakes and then frame a 2 x 4 to the stringline, making sure it is perfectly straight. (See Figure 6-1.) Overlap the 2 x 4 against the foundation

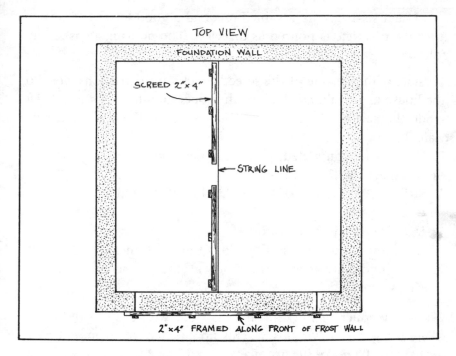

TOP VIEW

FOUNDATION WALL

SCREED 2"x 4"

STRING LINE

2"x4" FRAMED ALONG FRONT OF FROST WALL

Properly placed screed boards and stakes to section off pouring
Figure 6-2

wall on each side and stake it tightly to the corners of the foundation, as shown in the figure. Make sure the kicker stake is at a 45 degree angle. If the angle is less, the 2 x 4 framing could lift.

For a concrete floor wider than 15 feet, you must place a screed. This is done by running a stringline from the center of the back line of the floor to the center of the front 2 x 4 and then nailing a 16-foot 2 x 4 to three stakes driven not too deeply into the ground. (See Figure 6-2.) The screed 2 x 4 serves as a guide so that the floor can be poured as flat as possible with little wasted or overpoured concrete. It also helps in achieving the correct elevation. When one side is poured, remove the screed and pour the other half to meet the first side.

Use a six-bag, air-entrained mix and a good linseed oil sealer after the concrete is poured as directed. The pouring steps are as follows:

1. Start with one side of the screed. Back up the truck as close to the house as possible and shoot the concrete into a section. Then grade the section to the chalk line marks on the back and side walls.
2. Once this is completed, pull the screed and pour the other section to meet the first section poured.
3. Strike off the floor, or use a jitterbug if desired.
4. Bullfloat the floor.
5. Edge the front of the floor.
6. Cut joints to divide the floor into four sections.
7. Float and hard-trowel the floor at the proper time in the finishing process.

Many concrete workers finish a floor with a lip in the front where the garage door closes. This lip keeps water from entering the garage. To make the lip, place a 2 x 10 or 2 x 12 flush with the wood trim, a door width apart and flat on the surface of the concrete. Then push it down into the wet concrete ½ to ¾ inch to form the lip, and edge with an edger. Pull the board only when the concrete is set up enough so that the lip does not collapse. Broom the section on the outside of the lip to provide traction for vehicles entering the garage.

Basement Floors
Basement floors can be poured either directly after the foundation and necessary plumbing are installed or after the house is framed and finished. In either case, a basement floor sits directly on the inner edge of the footing. Stone is graded to the top of the footing throughout. Pour the floor level. (The only pitch is around the drain.) If the house is already framed, pour the concrete through the window well by using concrete chutes. It can be poured from

the different window wells or, using movable chutes, from one window well. In any case, once the concrete is placed and graded, use a jitterbug to level before bullfloating.

A basement is usually screeded between lolly columns, and each section individually worked in this manner. The columns that hold up the cross I-beam must be plumbed before pouring the concrete. Set up the basement by shooting in the corners based on the thickness of the floor above the footing. Then snap chalk lines as a guide for pouring and grading the concrete. Make sure that the concrete is in a workable consistency with a slump of approximately 4 to 5 inches. Two fundamental rules are broken when pouring a basement:

1. No expansion joint is used between floor and wall (concrete to concrete) and no control joints are put on the floor.
2. A basement floor gets a glass-smooth, hard-trowel finish for easier cleaning. Tile can be placed on its surface so water will flow easily to the floor drains, if it is necessary.

Don't worry if the sump pit is not the same height as the floor; it can be equal to or higher than the proposed floor. If it is higher it can be trimmed with an Exacto knife after the floor is finished and hard enough to walk on. The floor drain pipe also can be higher because it is usually adjusted to the level of the floor and topped with a drain cap. Slightly pitch a circular radius around the floor drain to allow water to drain easily. To create this radius, use a hand float in a circular motion in the area around the drain and keep constant pressure on the inside of the float.

For pouring a basement, a five-bag, 3000 psi concrete is sufficient, as the finished basement will not be subjected to exterior freeze-thaw cycles, salt and alkali action. Figure 6-3 shows how a basement floor is poured. Connect a 16-foot chute from the inside of the basement to the chute of the concrete truck (through the window well) and move the mix into the basement. Attach a chute

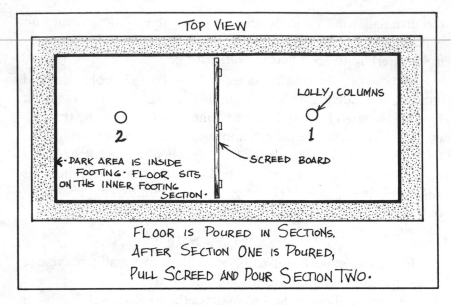

TOP VIEW

LOLLY COLUMNS

O
2

O
1

←·DARK AREA IS INSIDE
FOOTING· FLOOR SITS
ON THIS INNER FOOTING
SECTION·

SCREED BOARD

FLOOR IS POURED IN SECTIONS.
AFTER SECTION ONE IS POURED,
PULL SCREED AND POUR SECTION TWO.

Screed set up between lolly columns before pouring concrete
Figure 6-3

hanger to a wood joist in the basement to help move the concrete chute around. If a chute hanger is unavailable, move the inside chute by hand. With a large basement, you may have to move the truck and chutes to different window wells. If it is raining at the time a basement floor is being poured, make sure the sump pump is operative; if not, install a portable one with automatic operation.

Steps and Stairs

Proper stair construction requires an understanding of basic forming and finishing. Failures most often show up in blowouts and in improper finishing. Print specifications indicate tread width, riser height, and the length, width and type of stairs. Taking the preceding specifications, make a pre-fab form for small stairs from ¾-inch plywood as shown in Figure 6-4. Once made,

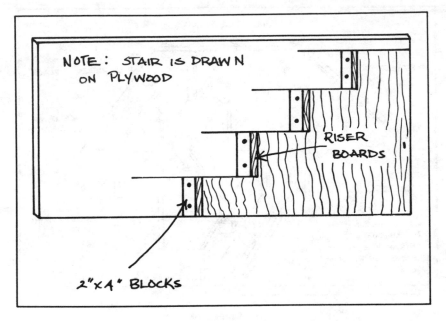

Plywood prefab form for small stairs
Figure 6-4

measure the width of the stairs, and frame both side forms as shown in Figure 6-5. Now cut the riser boards to proper length, fit them between forms, and nail them. When nailing the riser boards, leave the nail heads out so the risers can be stripped easily for finishing.

On larger stairs that are very long, leave the ¾-inch plywood side forms in one place and draw the exact stair design on the form. Nail 2 x 4 blocks on the side forms to keep the riser boards in line and straight. Here again, leave the nail heads exposed so the blocks can be pulled easily for finishing. On longer stairs, attach a middle brace made with 2 x 6 wedges to each stair to prevent sagging when the concrete is poured. (See Figure 6-5.) Secure the middle brace at the bottom ground elevation so the stairs will not shift when the concrete is poured. It is very important to brace the

75

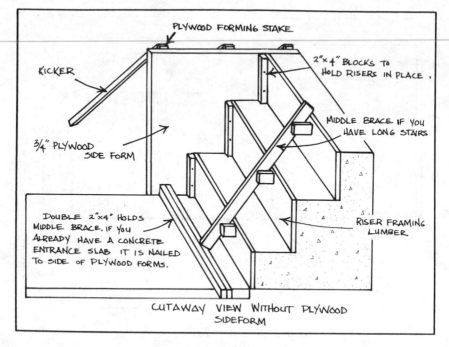

Formed concrete stairs
Figure 6-5

forms adequately. Stripping extra bracing is a lot easier than repairing a blowout caused by poor forming. Tips to remember when pouring are:

1. Use a low-slump but still workable concrete mix.
2. Start from the lower stair and work up.
3. Pour evenly and consistently and *never directly at one form section.*

The finishing tools needed are an edger, step tool, float, trowel, broom and safety groove edger (if required). You might prefer to make a B-6-12 curb butterfly tool to work the step. The advantage to making a custom step tool is that it reaches higher on the concrete riser and longer on the width of the tread.

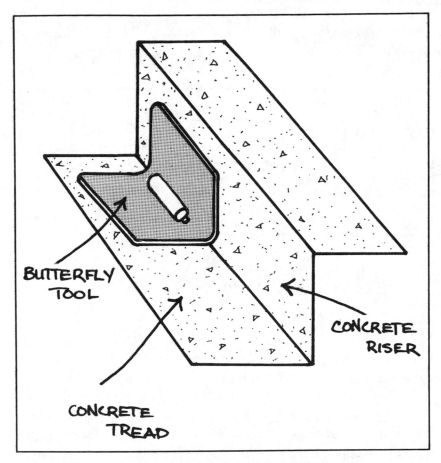

Work the butterfly back and forth on each tread and riser
Figure 6-6

Once poured, floated and edged, and when the cement is of a tacky consistency, yet strong enough so the concrete will not pull away or droop, strip the stairs. Work the butterfly back and forth on each tread and riser. The concrete on the face, step and origin point of the stair in Figure 6-6 is now ready for brooming. Brush the stairs just as you would a B-6-12 curb, by starting on the edge of the step and moving straight to the origin and up the face of the step. Once completed, allow the stairs to cure properly according to ASTM standards.

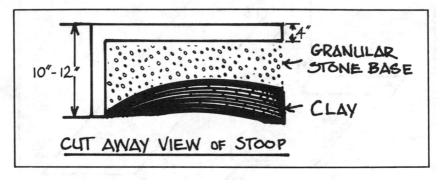

Side view of granular base and poured stoop
Figure 6-7

Stoops

The ideal way to pour a stoop is on a wing wall, pier or bordering foundation wall that goes below the frost line. This will ensure that the stoop will not drop or pull away from the house.

The normal range in step-downs from the largest front entry stoops to the small 3 foot x 18 inch pads is 6 or 7 inches. The surface of a stoop should always be below the interior floor grade to which it is applied. To find the starting elevation for the stoop, mark 6 or 7 inches down from the sill of the doorway. Pitch the stoop, no matter how small, ¼ to ⅛ inch per foot away from the building.

The lumber used to set up most stoops is 2 x 10's or 2 x 12's. These dimensions are used for appearance's sake, because stoops that have foundation wall perimeters have a visible face. For example, Figure 6-7 shows a 4-inch thick stoop around the outside face of a 10- to 12-inch-high stoop. The face of the stoop gives it the appearance of being 10 to 12 inches thick throughout when in

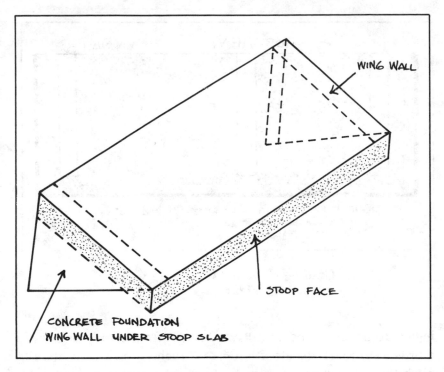

WING WALL

STOOP FACE

CONCRETE FOUNDATION
WING WALL UNDER STOOP SLAB

Stoop poured on a wing wall
Figure 6-8

reality, most stoops are only 4 inches thick. Figure 6-8 is a picture of a stoop poured on a wing wall. Notice how the face of the stoop is thick in the front and at the sides to hide the concrete wing wall.

In grading stone between wing walls, place the stone to the top of the wall all the way out to the end. (See Figure 6-9.) This reduces the amount of concrete required.

To frame small stoops, make a form with the 2 x 10's or 2 x 12's. Fit it to the location of the door (usually the stoop is the width of a door). Mark 6 or 7 inches down, place framing and stake it. Put kickers around the circumference, nail them to the frame stakes at a 45° angle, fill with stone (except around the outside perimeter of the stoop) and pour. Larger stoops require a str-

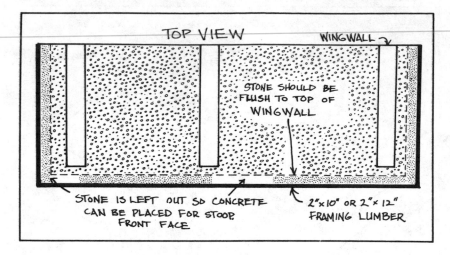

Grading stone between wing walls
Figure 6-9

ingline to be set up and 2 x 10's or 2 x 12's framed to it, as done in a driveway. Mark the elevation 6 or 7 inches down from the fill, pitch ¼ to ⅛ inch per foot, and set up stakes and stringline to this elevation. Start framing by using kickers on every upright stake. Fill with crushed limestone to the top of the wing walls, foundation wall extension or piers, making sure there is a good face for the stoop. Once the concrete is poured on all stoops, strike off, grade and bullfloat the stoop. Hand float and trowel at the proper time, and broom perpendicular to the house. Oil the forms on the inside face perimeter of the framing boards and strip the project the next day. On large stoops that are very wide, be sure to put control joints every 10 to 15 feet as needed.

7

Foundations, Detached Garages and Head Walls

A s the foundation is the starting point of every structure, it must be put in properly. Architect's prints, surveyor's measurements and excavation and grading should all be in order before any layout can take place. Once these items have been checked and their accuracy determined, begin the procedures for setting up to form.

Many forming systems are now available. Though expensive if commercially purchased, an average set of forms will afford you about 160-185 pours if proper care, including refacing, is taken. A relatively new forming product system uses styrofoam panels that can be put together on a jobsite so that panel forms are eliminated. Yet for all the advantages of the various forming systems, do not neglect to consider the old-fashioned system of plywood. The plywood system has many jobsite advantages and each individual contractor develops what works best for his particular needs.

Though each of the forming systems I have mentioned is constructed differently, all forming principles still apply, and layout
and framing remain basically the same.

A concrete contractor should be concerned only with pouring, so let a qualified excavator dig the holes. Start by setting up
the transit close to the excavated hole. Check to make sure that all
property pins are correctly in place and determine two base lines
from pin to pin. This gives the information needed to set up the
batterboards. Batterboards intersect at 90 degree angles with the
foundation wall lines. Erect them far enough away from the hole
so interference with forming traffic does not occur. Figure 7-1 is
an overhead view of this procedure. The length of batterboard
varies, but they are usually 3 to 4 feet long and must be sturdy.
Pound three stakes into the ground to hold up each batterboard.

Now establish a benchmark, the location of which is usually
included in the grading plan prints. This reference point elevation
is frequently a fire hydrant bonnet bolt marked with an ''X.'' To
determine the top of the foundation wall, check two fire hydrant
benchmark points against each other. Note the benchmark elevation reading through the instrument and subtract the grading plan
foundation elevation to see how much change is needed in elevation from the benchmark. This will give the exact height of the
foundation walls. Mark this elevation on the stakes and transfer
the proper measurements from the base line before erecting the
batterboards to the height of the foundation walls.

Set up four corner batterboards and erect them to the proper
top wall elevations. Drop a plumb bob to determine the building
line corners of the outside walls, and place a nail in a stake top to
mark them. Measure out from these corners and use a stringline in
a box-type setup to determine the outside footing lines. Make the
corners square by measuring diagonally from corner to corner.
The rule of thumb is to make the footing twice the width of the
wall, so an 8-inch wall gets a 16-inch footing and a 10-inch wall

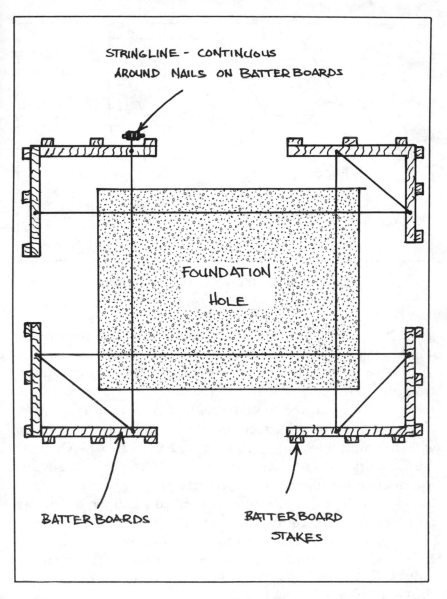

Setting batterboards
Figure 7-1

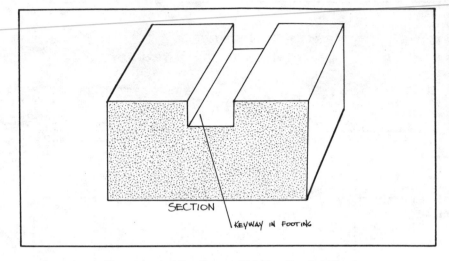

SECTION

KEYWAY IN FOOTING

Keyway to tie the wall into the footing
Figure 7-2

gets a 20-inch footing. If constructing an 8-inch wall, measure out 4 inches to place the outside face of a 16-inch-wide footing.

Frame the outside by first measuring the building lines down from the top of the wall to determine proper elevation of the footings. Note that the footings must be placed on undisturbed ground. Check for print specs on footing thickness, and frame with the proper lumber, usually a 2 x 10, 2 x 12 or 2 x 14. Once the outside is framed with stakes and kickers, measure in and frame the inside of the footing. Use spreaders together with framing to keep the footings set at the proper specified width. Spreaders can be 2-foot stakes nailed to the top of the footing boards.

Keying the footing to the wall is an alternative to print specs for commercial buildings that require rebar from the footing to the wall. A keyway is a 2 x 4 located in the center of the footing, floated in flush with the top and then pulled out at the proper time. This makes a notch in the footing as shown in Figure 7-2. This procedure makes it possible to tie the wall into the footing.

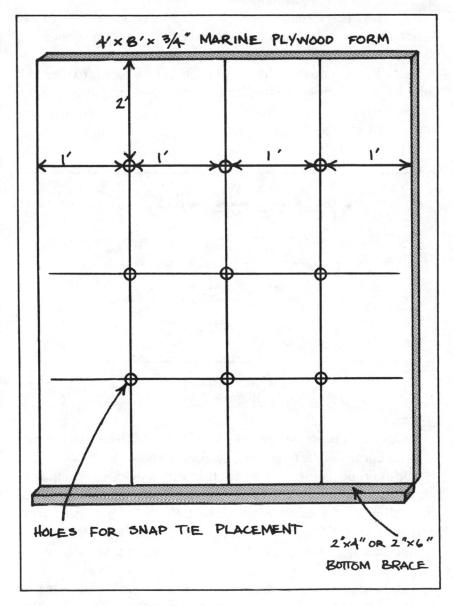

Placement of holes for snap ties
Figure 7-3

The plywood system of forming can be set up by using ¾ inch 4-foot x 8-foot marine plywood. Metal bordering with connecting hardware is available or you can substitute 2 by 6's as bottom runners. If fillers are needed, divide a full sheet into panels to meet required specifications. In either case, stack panels and drill holes at 2-foot intervals in 8-foot dimensions and 1-foot intervals in 4-foot dimensions to accommodate the snap tie as shown in Figure 7-3. Figure 7-4 shows a typical snap tie.

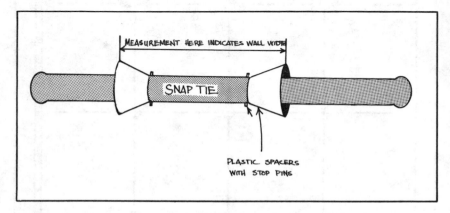

Typical snap tie
Figure 7-4

Once the proper number of panels for the job has been made, lay the forms against the four walls of the hole. Once again, establish proper corners on the footing with a plumb bob and snap chalk lines to indicate the outside of the foundation wall. Check the wall again for squareness on the footing by measuring from corner to corner. If the measurements are equal in length, the wall is square on the footing and panels can be erected. Usually one or two workers set up the outside wall and two more workers follow and place the snap ties in the pre-drilled holes. Set the outside wall to the chalk line and nail the base runner with concrete nails either to the footing itself, or to the footing's 2 x 10, 2 x 12, or 2 x 14 boards. Once the panels are nailed to the chalk line, assemble the form clamps on the snap ties and wedge a 2 x 4 brace or waler between the outside panel and a ledge on each clamp. (See Figure 7-5.) The proper setup of both the inside and the outside walls with walers is shown in Figure 7-6.

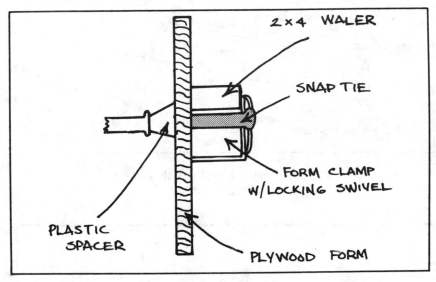

Side view of assembled snap tie, form and clamp
Figure 7-5

After the walls are set up, erect doubled 2 x 4's vertically for additional bracing and nail each to the 3 or 4 walers in each 8-foot section. In addition, nail a double 2 x 4 vertical brace at each end of a 4 x 8 panel to cover the break between panels. When the vertical braces and walers are in place, install long 45 degree angle 2 by 4 braces to further steady the wall at the top and bottom. If the panels do not come out evenly in framing, use a filler made to exact measurement for finishing. To complete the framing, nail the walers together to strengthen the corners.

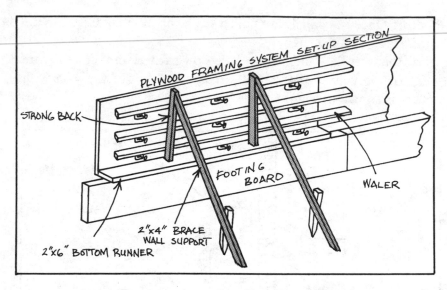

Proper setup of walls with walers
Figure 7-6

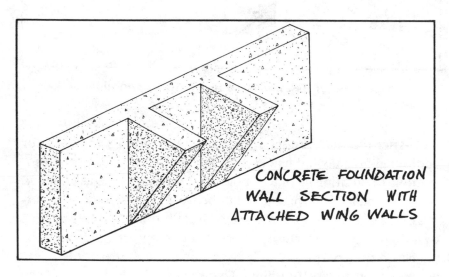

Typical wing wall
Figure 7-7

Many specifications indicate that wing walls are to be built into the structure. A wing is an extended short wall foundation ledge which is used as a base for stoops, porches or landings. It is a nice safeguard against settling problems that can occur with bordering outside flatwork. Figure 7-7 shows a typical wing wall. It's a good practice to check and recheck the framing and bracing and look over the print for any basement window wells or escape wells that must be placed. Place any window wells or escape wells 6 inches below the top of the wall and tack them with nails inside the forms. For a sound foundation construction, wait approximately one to three days before erecting the wall on the footing, then strip the wall within one week. Do not strip a green wall.

Another forming method that is noteworthy is the styrofoam system. Everything in practice from the layout to the footing remains the same as the plywood system; the only difference is in the wall paneling system. Foam forms are basically big building blocks and are set on the footing by interlocking panels. Blocks are put together at the jobsite. Both the contractor and the owner benefit from the styrofoam system. The contractor saves because he does not have to invest in reusable forms, has no need for a form truck, and requires no labor to haul in and strip forms. The owner gets an insulated wall which is energy efficient.

Foundation Additions
As the costs of both moving and newly constructed homes are constantly increasing, additions and added-on garages are becoming a major part of the business of a small concrete contractor.

Layout for any addition begins by marking the corners of the building with stakes and squaring the addition to the existing structure. Pound a stake next to the existing foundation 2 to 3 feet back from the end of the wall. Attach a stringline to the stake on the inside next to the wall, then measure the width of the front of the proposed structure from the foundation. At this point, locate a corner stake temporarily. Now stretch the stringline to this stake and hold the string out at an angle from the foundation wall of the structure, then move it slowly in until the string touches the wall along the existing foundation. Move that stake to the stringline to give the front building line and corner as shown in Figure 7-8.

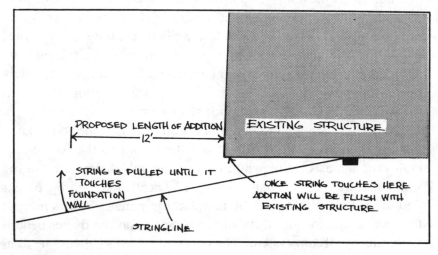

Squaring up proposed foundation with existing foundation
Figure 7-8

Repeat the procedure on the back side of the building. Square the building by measuring the diagonals from outside corner to outside corner. If these measurements are equal, the building is square and batterboards can be set up. Batterboards are corner boards that are staked away from the actual line of digging, but still indicate the exact building lines. This is shown in Figure 7-9. Note that the strings are extended out to the batterboard so that the hole can be dug without taking down or disturbing any important elevations or building lines. If trenching is being done, foundation walls or a full foundation can now be dug without upsetting any measurements.

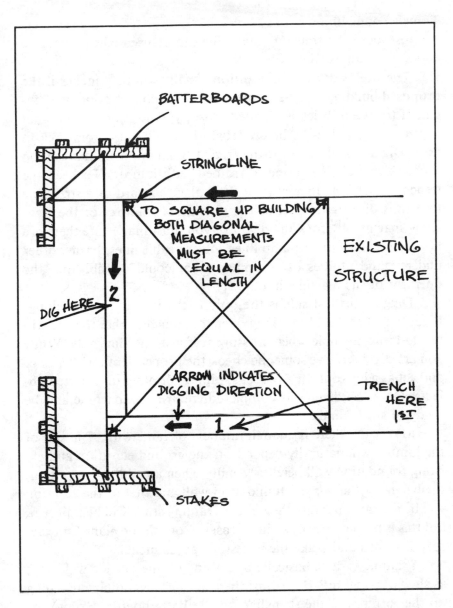

Batterboard setup and trenching for attached addition
Figure 7-9

Trench Foundations

In most locales, a trenching machine can be used when a room is added onto an existing structure.

The wall width of the addition usually will be 8 inches if the proposed building is to be framed with 2 x 4 construction or 10 inches if the use of brick is planned. Also, you must determine what depth is required to be below frost. This will vary from area to area. Dig a straight line for the trench foundation so that the top exposed wall can be properly framed with lumber. To keep the trenching machine in a straight line, place a board or a stringline that runs directly parallel to the outside of the tires of the trenching machine. Before any digging, you must make sure there are no electric lines, sump pump lines, phone cable lines, or any other underground utilities or equipment that could possibly put the operator or the machine in danger.

Digging the last side is the only tricky part about trenching. Place 2 x 14 or 2 x 12 planking over the trench so that the tire will not fall into the hole when crossing to finish the final side. While you are digging, have someone check the proper depth with a ruler and act as a lookout for any rocks, piping, or other unexpected objects. Once the digging is completed, return the machine and be ready to start framing.

Make sure there is enough lumber to frame with. The size of the lumber will naturally depend on the ground elevation and existing foundation wall height. Usually when an addition is put on, the height of the existing foundation wall and that of the addition will be the same because the lumber framing studs will be uniform, but this is not necessarily so in all cases. Consult the plans for exact specifications and make the necessary adjustments.

Framing is done basically by nailing up the outside wall first, as shown in Figure 7-10. Adjust the framing elevation by matching up the forms with the stringline on the batterboards. Extend the framing board about 1 foot past the hole so that a stake can be

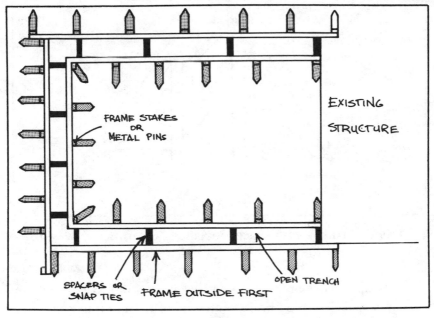

Framing is done by nailing up the outside wall first
Figure 7-10

pounded into the ground without disturbing or caving in the trench. Now nail the framing board to the height of the stringline. The other ends of the boards are also extended approximately 1 foot past the existing foundation. This is done so stakes can be put into the ground away from the trench. Set the height as before, cut and connect the boards and nail to each side. This results in an "H" shape. After this is done, put the supporting stakes and kickers along the complete perimeter, straightening the framing board to the stringline. Figure 7-11 shows framing stakes. Frame the perimeter boards in a triangular brace from top to bottom and connect them to a stake pounded in approximately 2 feet from the hole. This is the only way to effectively frame the boards without caving in the hole.

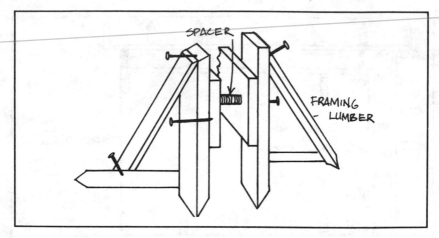

Framing stakes with spacer installed for proper wall width
Figure 7-11

To help achieve proper wall width spacing, put wood pieces in the center 8 inches wide or 10 inches wide for brick as shown in Figure 7-11. The wood spacers are usually 2 x 2 cuts and are knocked out when the concrete is filled to the top. Always put as much bracing as needed to make the form solid, and anything in question should get extra bracing. It is consequently much less expensive to use more 2 x 4 lumber for bracing than to have the form blow out and ruin the job. Once all framing and bracing is completed, doublecheck the elevations and squareness of the building. If all is proper, pour the concrete, remembering to keep the truck and concrete chute away from the framing and bracing. Also, try to pour as much of the wall from one spot as possible. Pour slowly and evenly into the hole and not at the framing board. To aid in this, hold a shovel perpendicular to the end of the chute to direct the rushing concrete directly into the hole. When pouring, have someone work the concrete in an up and down motion with a 2 x 2 or 2 x 4 to consolidate it and eliminate voids.

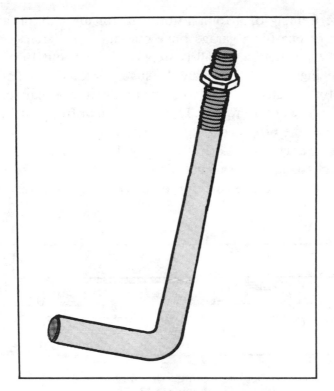

Foundation bolt
Figure 7-12

When pouring is completed, hand-float the concrete wall top and get ready to place the *foundation bolts*. A drawing of a foundation bolt is shown in Figure 7-12. If you are using a 2 x 4 base plate, embed the bolts in the concrete, leaving about 2½ inches exposed in addition to enough space to accommodate another 2 x 4 and a nut over the bolt. Place them into the surface about 1¾ inches in from the outside of the wall. This gives enough room to center the 2 x 4 on the base plate and enables it to sheath the side and the wall, finishing just at the outer edge.

The framing of a foundation wall for an addition and the framing of one for a garage have one major difference. At the front of the garage, the wall is stepped down about 10 inches as shown in Figure 7-13. This provides space for the garage door. The reason for this step-down is to permit an entrance that correlates with the grade of the ground. The step-down or frost wall becomes the base of the proposed floor in the front of the garage. When pouring a garage floor, keep that floor 1½ to 3 inches below the rear of the foundation wall and pitched ¼ to ⅛ inch per foot to the front of the garage. The floor of a room addition requires no pitch.

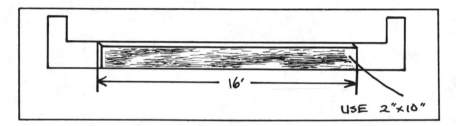

Framing a garage foundation wall
Figure 7-13

Detached Garages

A garage attached to an existing home is far preferable to a detached garage. Some lot widths, however, eliminate the possibility of an attached garage because of code specifications on widths between lot lines, houses and easements.

Detached garages are basically garage slabs with a grade beam foundation. This is a foundation that is incorporated monolithically as part of the garage floor. The depth of this 8-inch wide wall is usually 18 inches to 36 inches. Pouring a detached garage slab is a lot less expensive than an attached one because it requires less concrete and fewer workers.

With the plan and final floor elevation of the garage already known, begin setting up where the garage will be located and square the building off as before. Depending upon the height of the finished floor, choose the lumber and begin framing the outside perimeter of the forms to the stringline. Don't worry about having a gap under the stringline because fill can be used to overcome this problem. At this point, depending upon the depth required, dig 18 to 25 inches deep, or if deeper, trench the wall with a trenching machine. If trenching, set up and dig with the same procedure shown in the additions section of this chapter. After digging out the 8- to 10-inch wide trench, set up and frame the inside boards. This time use a 2 by 6 which leaves a 6-inch gap from the ground grade to the bottom of the 2 x 6. This gives enough depth to put in a 2-inch stone base and still have 4 inches of space left for the concrete. After framing the inside 6 inches above grade and flush with the outside top 2 x 12, pour the floor and then pour the wall of the foundation immediately afterward.

Head Walls
By many standards, head walls seem a very insignificant part of the forming system picture, but they are necessary. If you are working on large paving and sewer jobs, make sure you thoroughly understand the construction of head walls because they are included in the construction of many pipe runs or containments. Many contractors who obtain large concrete jobs from paving outfits and generals have to subcontract head walls because they lack the knowledge to construct them, even though they are highly profitable. Head walls are bid without any excavation costs. Excavation is usually taken care of by the general contractor.

Begin by examining a print and specifications for the type of wall to be constructed. Stake out the floor of the head wall if required. Once staked and framed in accordance with the print design, place and compact the stone base. Cut and lay out steel rebar or wire mesh next if required. Many states and municipalities use rebar according to a standard plan that incorporates tying in the floor and walls in a very efficient manner.

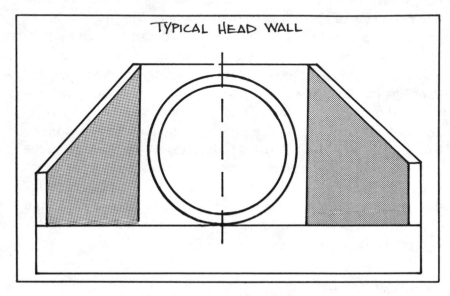

Framing a basic head wall design
Figure 7-14

After pouring and finishing the floor, bring in lumber, stakes and forms for framing the walls. Figure 7-14 gives an idea of how to frame a basic head wall design. The plywood needed can be either ¾-inch 4 x 8 sheets or any commercial forming system. If using plywood, consider using an efficient system of clamps to be used with snap ties called *form clamps*. This type of clamp simply connects to a snap tie and holds the plywood at uniform width according to the plan.

Begin by snapping a line on the floor to correlate with the proper placement and alignment of the forms. Once snapped, place a 4 x 8 sheet with a 2 x 4 bottom onto the floor slab and pound

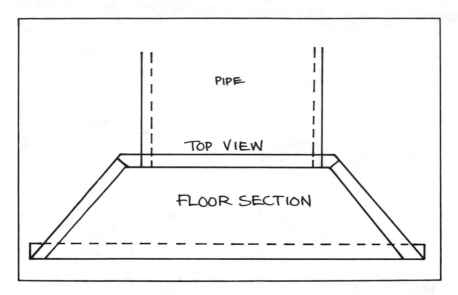

**Pipe entering mid wall section onto floor that the sidewalls sit on
Figure 7-15**

stakes in the ground next to the slab. Before any plywood is erected, drill a total of six holes in the 4 x 8 sheet, each hole one inch in from each side. This equally divides the plywood. Hole sizes must correspond with the diameter of the snap tie so the snap tie will be able to fit through the plywood. Put up one side and nail the stakes into the 2 x 4 attached to the bottom of the form. Now, put the snap ties into each hole and erect the other side by placing the snap ties in the opposite holes and securing the ends with form clamps on all snap tie ends. Then pound 2 x 4's or walers into the face of the form clamps. This will stiffen the ¾-inch plywood forms and make the system capable of containing concrete. Once the walls are erected and braced, cut a plywood piece to extend over the face of the pipe and attach to both walls. Then place a single row of snap ties to hold the top portion of the wall directly over the pipe. Secure the forms to the side walls, erect braces at 45° angles, and nail them to stakes in the ground.

99

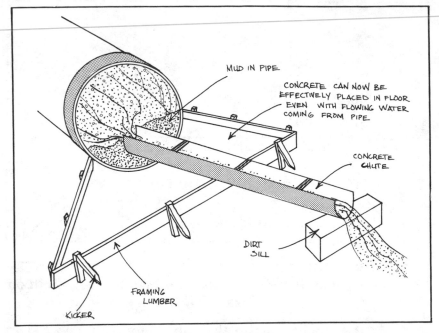

**Method of pouring floor of head wall when water is already
flowing through pipe
Figure 7-16**

If a lot of rebar is put in the walls, pour concrete stiff
(2½-inch slump) and then vibrate in lifts of 32 inches. With the
concrete in place, let the top of the wall cure and then strip it at the
proper time. If the forms are to go higher than 4 feet, simply stack
another 4 x 8 sheet with 2 x 4 bottom on top of the lower form, nail
walers to strong backs, and put up braces.

Remember these points:

1. Always maintain proper bracing, especially with any filler
pieces.

2. When the wall is stripped, break off the snap ties and patch the
holes with recommended quality patching materials.

3. If water is coming out of the pipe, put a 16-foot concrete chute in the base of the pipe and surround the chute with mud. Then erect a sill for the other end of the chute to sit on so water will be carried over the floor as shown in Figure 7-16.

8

Finishing Flatwork

The process of finishing in flatwork concrete construction includes several steps, each of which is equally important to the quality of the finished product. Striking off is the first step in the finishing process after the concrete has been properly placed. Place a 2 x 4 on the surface of the concrete and move in a combination side-to-side and forward motion. This pushes down the aggregate in the concrete while bringing up the cream, and also levels the surface in preparation for bullfloating.

Bullfloating further refines the smoothing process, flattens the concrete and brings the necessary cream to the top of the mix. To accomplish these all in one motion, adjust the bullfloat at an approximate 30° angle as shown in Figure 8-1. Work the bullfloat by pushing it smoothly and evenly on its near edge over a section of concrete, and slightly raising it on its far edge while pulling it back again (see Figure 8-2). If a good strike off does not result,

Keep the bullfloat at a 30° angle, working back and forth over the concrete
Figure 8-1

Pull the bullfloat on its far edge and push it on its near edge
Figure 8-2

move the bullfloat in shortened motions to knock down any stones still apparent on the surface. Bullfloats are available in widths from 24 inches to 4 feet. The five basic rules to keep in mind when using this tool are as follows:

1. Bullfloat as soon as the concrete is struck off.
2. Use the bullfloat to fill in low spots by sprinkling on material and bullfloating the concrete flat; also use it to knock down high spots.
3. Always overlap each stroke until the entire width of the job is completed.
4. Keep the bullfloat as clean as possible and wet it down before placing it on the concrete.
5. The bullfloat can indicate if the concrete is flat by not leaving any ridges after overlapping. If a line or slight ridge is left, raise or lower the concrete accordingly.

After bullfloating, form a radius curve on the outer border of the concrete, or edge by passing an edging tool. This tool is used with the same motion as the bullfloat. Place an edge in the concrete with the near end of the tool and pull it back with the far edge down. (See Figure 8-3.) Basic points to remember are:

1. Keep equal downward pressure on the top of the edging tool at all times.
2. Hand float the area around the perimeter of the boards before edging.
3. Use a smooth back-and-forth motion to produce fine edges.
4. Keep the 2 x 4 form edge tops clean by scraping excess concrete off with the flat or curved part of the edger.

If the concrete is too wet to run good edges, wait a little while until the edger makes clean, concise edges.

The next step is called *cutting the joints.* This can be done from 20 to 45 minutes after edging, depending upon the temperature and how quickly the concrete sets up. The best time to

Edging the concrete
Figure 8-3

cut joints is usually when the concrete is hard enough to support the kneeboards, yet pliable enough to allow easy cutting of the joints. The tool, called a *jointer*, creates a·cut in the concrete. The depth of a joint usually ranges from ¼ to 1 inch. The purpose of a control joint is to create a weakness in the concrete which will cause any cracks, if they occur, to take place at that joint. The jointer is worked as follows:

Place a straight 2 x 4 (usually 16 feet) across the surface of the concrete exactly where the joint is needed. The 2 x 4 will act as a guide. Now, run the jointer edge along the 2 x 4 with a pushing motion, the near edge down and the far edge slightly above the surface of the concrete. A pulling-back motion with the near edge slightly up and the far edge down will finish the joint. Run the jointer back and forth until a neat joint is achieved, as shown in Figure 8-4. On larger jobs, use the jointer together with a bullfloat attachment and handles when tooled joints are required. Basic rules to remember are:

1. Don't press too hard with the jointer unless the concrete is hard.
2. Always keep the near edge of the jointer down when pushing and up when pulling.
3. After cutting the joint, remove the 2 x 4 carefully by wiggling side to side a little until the suction is released.
4. Cover up the kneeboard marks by floating. Also, float out the jointer marks.
5. Rule of thumb on surfaces—place a control joint for every 10 to 12 feet of concrete. If the concrete is 20 feet wide (as in a driveway), use an expansion joint filler.
6. Always check print specifications for location of control joints in the concrete.

Once jointing is completed, the final steps in the finishing process are troweling and broom finishing (if so desired). Troweling is done by moving the trowel with the arm at a slight angle to the concrete while advancing in a semi-circular motion as shown in Figure 8-5. This movement will produce a hard smooth finish. Important points to remember are:

1. Make sure the four corners of the trowel are bent slightly up to avoid catching and marking the surface of the concrete.
2. Keep the trowel at a slight angle (less than 1 degree). A larger angle will produce ripples in the cement.
3. Remember to keep even pressure on the trowel. Leaning on one end or the other will cause dips and trowel marks.
4. Begin troweling when the concrete is tacky to the touch.
5. Reach and trowel the outside perimeter of the work first. This allows you to make one troweling pass in the center.

Run the jointer back and forth until a neat joint is achieved
Figure 8-4

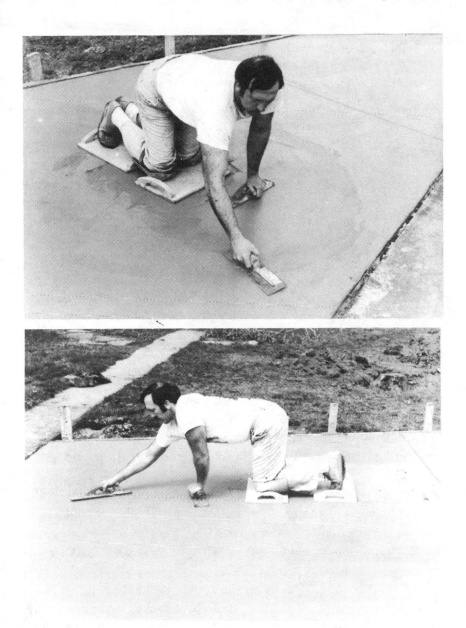

Keep your arm at a slight angle; trowel in a semi-circular motion
Figure 8-5

6. Never trowel over a joint. Try to parallel the joint as closely as possible. Do this also with the edges.

7. Work as quickly and as smoothly as possible. You can always make a second pass if needed.

Used together with the trowel is the *hand float*. The hand float is used in the same manner as the trowel. Points to remember are:

1. Use the float with the trowel to level and smooth the concrete in the finishing process.

2. The float can be used to complete the final finish of the work if a hand trowel finish is not desired.

3. Work parallel to the joints and never over them.

The cement is now ready for a *broom finish*, if one is desired. This is done by pulling a broom lightly across the surface of the work to produce a shallow grooved finish. The broom can be pulled straight or curved to create different effects. Points to keep in mind are as follows:

1. Pull the broom lightly and evenly over the work.

2. Always keep the broom clean.

3. After one pull across the concrete, dip the broom into water, shake it, and make the second pull, overlapping ¼ inch over the first pull.

4. Try to broom in the easiest direction possible, preferably parallel with the joints. (See Figure 8-6). If not possible, then broom perpendicular to the work.

5. Bring the broom as close as possible to the joints, but not over them.

6. When it is *slightly* tacky to the touch, concrete will produce a light broom finish. When the concrete is tacky, a medium finish will result and when it is wet and tacky to the touch, a rough broom finish will be produced.

When possible, broom parallel to the joints
Figure 8-6

7. A light-to-medium broom finish is fine for patios. A medium broom finish is appropriate for sidewalks. A heavy medium-to-rough broom finish is good for driveways.

Once brooming is completed, allow enough time for the concrete to become hard enough so that it cannot be marred. Then, if desired, apply a concrete *curing compound* and *sealer*. A sealer is basically a chemical compound that can be applied to the surface of the concrete to protect it from spills of oil and grease and various alkali action. Sealer also helps in the curing process by slowing down the hydration process, making the concrete stronger and more durable. This is invaluable especially if the job is in an area subject to constant freeze-thaw cycles each year. Another feature of a sealer is that it dustproofs the surface, which can be especially important for interior floors such as basement and garage floors. An excellent sealer made by the Sonneborn Company is called *Cure-N-Seal.*

A different type of sealer that is especially effective against alkali action, such as salt, is *linseed oil sealer.* This is usually applied after 28 days of curing and will protect concrete against salt action for approximately 200 to 300 freeze-thaw cycles. *Tri-Dar* made by Darling & Co. is an excellent linseed oil sealer. It is applied like Cure-N-Seal. It can be rolled on or sprayed on. For both sealers mentioned, traffic should be kept off for approximately 24 hours. The sealer can be reapplied after effectiveness has passed.

9
Weather Protection

There is no doubt that weather affects us all to some degree. In the construction trade, you know that each working day, you must take the weather into consideration when planning the day's activities. For the proper placement of good quality concrete, the weather should not be too hot or too cold. Ideally the temperature should remain constant for the 28-day curing period.

Concrete, if placed when the temperature is too hot, tends to harden too quickly, causing improper curing. At the other extreme, placing concrete in below-freezing temperatures and not handling properly can also cause problems. At the time a job is ready to pour, check the weather and evaluate the conditions prior to delivery.

The ideal day for placing concrete would be a dark, windless, humid day with a temperature in the low 60's. These precise weather conditions are rare in any climate and are almost nonexistent in some areas. It would be impossible to restrict the placement of all concrete to days when the weather conditions were most favorable. Although extremes of temperature and other weather conditions may add to the difficulty of placing, it is possible, with precautions and planning, to achieve excellent results in the heat of summer or in winter's cold.

Weather conditions, of course, are not totally foreseeable, such as when a sudden rain storm appears or a below-freezing night silently creeps up. This chapter explains the various problems that can result from different types of weather, how to prepare for them, and possible solutions to the problems.

Rain

Rain is probably the number-one enemy of exterior concrete while curing. If it starts to rain immediately after the cement has been placed, cover the work with Visqueen and wait until it stops raining. Then remove the plastic and continue finishing. If the rain is expected to continue, erect a covering to allow you to finish the job. The best shelter is a lean-to covering. Lean 2 x 4's against an existing structure and nail plastic over the boards. This should give enough room to finish the concrete. For covering sidewalks, nail 2 x 4's on the existing stakes. Lift the sides of the plastic to work under it as you go. Be prepared for bad weather by having enough plastic and 2 x 4's on hand at all times.

Extremely Hot Weather

Concrete at 60°F will set up in about 2½ hours and will be completely firm in about 6 hours. Concrete placed at 100°F may set up in 45 minutes and be completely firm in 3 hours or less. This accelerated setting, of course, greatly increases the difficulty of finishing the 100°F concrete. The chances of cold joints developing during placement are greatly increased because there is not enough time to strike off, float, and trowel the surface.

For each 10°F rise in temperature, an additional 7 lbs. per

cubic yard of water may be required to produce the same slump. If additional water is added to the mix, an additional measure of cement must also be added to maintain the water-cement ratio. If this is not done, strength and durability will be impaired. The addition of more water can also result in greater volume change in the concrete. The addition of the extra cement will, however, further increase the total heat of hydration. As concrete temperatures increase, rate of slump loss increases and the water requirement increases sharply (see Figure 9-1).

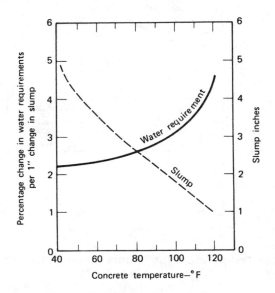

The effect of concrete temperature on slump and on water required to change slump
Figure 9-1

Concrete placed during hot weather without precautionary measures being taken is often less resistant to freeze-thaw and wet-dry cycles. Figure 9-2 shows the effect of high temperatures on concrete compressive strength at various ages.

Immediate remedies to the problems encountered when plac-

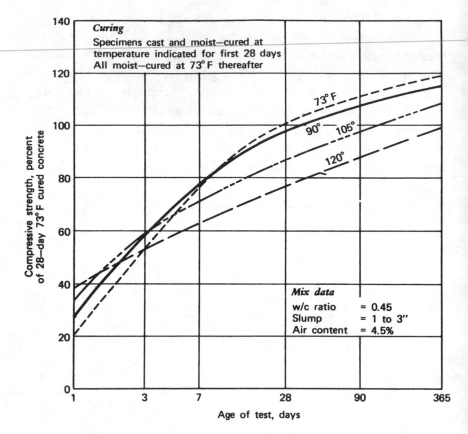

Effect of high temperatures on concrete compressive strength at various stages
Figure 9-2

ing concrete in hot weather are few. One is to apply a curing compound immediately after pouring. Follow up by moist curing according to ACI standards. Moist curing involves placing burlap or membrane paper over the entire surface of the concrete and keeping the material wet. The effect of moist curing on concrete compressive strength is shown in Figure 9-3.

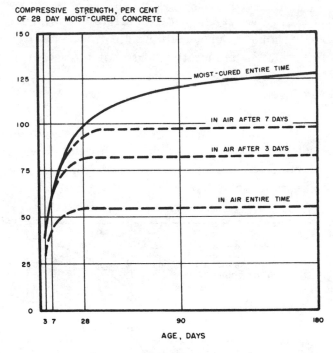

COMPRESSIVE STRENGTH, PER CENT
OF 28 DAY MOIST-CURED CONCRETE

Effect of moist curing on concrete compressive strength
Figure 9-3

Windy, Humid Days

On windy, humid days, a condition can develop called *crazing*. It is recognizable by the cracked appearance of the surface of the concrete, visible mostly when the concrete is wet. Basically, this is caused by the surface of the concrete shrinking at a faster rate than the rest of the mixture. Crazing is especially prevalent with the higher-psi bag mix cements. Immediate and proper moist-curing helps alleviate this problem.

Freezing

Plastic concrete must not freeze. The freezing temperature of concrete is near the freezing point of water (32°F). When the mix

117

water forms into ice crystals, the concrete is considered frozen. The concrete should be protected until hydration has used enough of the free water to reduce the degree of saturation and create enough strength to withstand stresses.

Cold concrete has slower hydration rates. At 40°F, it takes longer to develop a given strength than it does at 73°F. When subsequent construction depends on a quick strength gain in the concrete, the delays caused by cold weather interfere. Figure 9-4 shows the effect of low temperatures on concrete compressive strength at various ages.

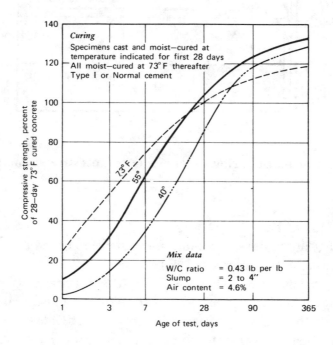

Effect of low temperatures on concrete compressive strength
Figure 9-4

If the concrete freezes before a compressive strength of approximately 500 psi has been developed, considerable damage is done. At best, the piece will have a damaged surface and will never

118

attain watertightness; at worst, the freezing will cause such heaving and structural faults that it will lack strength and deteriorate. The critical point at which this damage is apt to occur is determined by the design of the mix, the type of cement used, whether or not the concrete is air-entrained, and the degree of hydration reached.

The use of as low-slump concrete as possible is recommended. The less mix water there is to freeze, the sooner a safe strength will be attained.

In cold weather the temperature variations from top to bottom of a slab can lead to thermal stresses within the slab that can cause cracking.

The first 24 to 48 hours after placing are the most critical for concrete. Once the level of compressive strength has reached 500 psi, it can survive several freeze-thaw cycles. Favorable curing conditions with good protection must continue, however, for as long as possible.

Never pour concrete when you know that frost is in the ground. If you have already poured the work and freezing is forecast for that night, cover the work with burlap or Visqueen and then put approximately 6 to 8 inches of straw over the entire surface of the work to hold in the heat generated during the curing process. Leave this on for as long as possible, even though it can actually be pulled up once the concrete has reached a 500-psi strength tolerance. Make sure that the surface of the concrete is dry enough so that the covering does not damage or mar the top.

If you must pour during freezing conditions, use calcium chloride as a drying agent at a maximum of two percent of the total mixture. First check to see if specifications allow the use of chloride. It is a good idea to have the concrete company add liquid chloride at the plant. The chloride speeds up the hydration process so concrete can be placed and worked faster in more favorable conditions before freezing occurs. With any job, there always remains the element of risk, but with proper knowledge and precautions, many potential hazards can be overcome.

10
Custom Concrete

Because of its new and varied uses, custom concrete is now used by many architects and designers. Residential concrete contractors are also finding an increasing number of individuals who want unusual concrete flatwork in their homes. Consequently, being able to work with custom concrete can open up a new realm of business opportunity.

Exposed Aggregate
One popular style of custom concrete is exposed aggregate. To attain this look a concrete mix is used that contains rounded, variably-colored stones in addition to cement, sand and water. The concrete is mixed and poured in the usual fashion. When the concrete has set to a consistency that will not be too deeply marred, broom the surface, using a broom with a water hose connection. This broom sprays a continuous stream of water on the surface of

the concrete, exposing the aggregate and highlighting the colored stone. Figures 10-1A through 10-1D show the process used to place exposed aggregate.

Continue brushing the surface until the colored aggregate is uniformly exposed and there is no trace of cement paste on the surface of the stones.

Apply a good, clear non-yellowing sealer to the concrete. This will enable it to withstand freezing and thawing. Sealing also protects against corrosive elements such as salt, and helps ensure longevity and durability.

Size of the aggregate can vary from pea gravel to fist-size rocks, depending on the desired texture. Recommended sizes range from 3/8 to 3/4 inch. Aggregate less than 1/4 inch in diameter should be avoided. Either natural or white portland cements, with or without color oxides, can be used to augment the aggregate color. For many jobs, natural portland cement with locally available aggregate is acceptable. Native gravels are usually found in a range of browns. Some of the more attractive surfaces are obtained with aggregates containing a large percentage of black particles.

Colored Concrete

The art of coloring concrete has many professional applications. There are two ways to color concrete. The first, and by far the easiest and safest, is to use integrally-colored concrete. This is concrete that has been colored by exact proportion and weight with a powdered pigment at the factory. It will cost a little more, but the proportions will be exact and the color even.

Colored concrete must be used in a different manner than standard mix. The stone base must be firmly compacted, accurately and uniformly pitched, and thoroughly moistened. It is preferable to start the previous night and dampen the stone well so that the concrete will dry uniformly. A 4-inch slump is ideal, but a 5-inch slump is acceptable for hotter weather.

Distribute the selected, decorative aggregate to cover the entire surface evenly
Figure 10-1A

Embed it by patting it with a darby
Figure 10-1B

Darby the surface to completely cover the aggregate with grout
Figure 10-1C

Expose the aggregate by simultaneously brushing and hosing with
water
Figure 10-1D

After troweling, brush with a light to medium broom to enhance the color of the concrete. A further way to deepen the color is to use a good quality, clear sealer. A non-oil-based sealer like Sonneborn's *Cure-N-Seal* is ideal for this purpose.

The second method of coloring concrete is to use a dry pigmented powder. This method is more difficult, and is preferable only if deep color and wear resistance are of the utmost importance. Sieve the powdered mixture onto the concrete, using your fingers or a sifter, and trowel simultaneously with a wet wooden trowel. Be careful not to over-finish the surface, as this can produce an uneven color. If inferior pigments are used, discoloration and fading from the sun can result.

Stamped Concrete
Concrete stamping is a new technique which requires special tools and procedures to produce various designs and patterns in concrete flatwork.

When doing a stamped-concrete job, carefully plan the layout of the design. Before forming, make sure that the length and width of the area to be stamped conform to the particular stamping tool you are using, so that you will create a squared and even design. Reinforcement or expansion should be used just as in the normal preparation of concrete flatwork, even though stamping will afford better crack concealment.

First, lay a 4-mil visqueen flat on the concrete's surface, pulling all sides of the plastic evenly to take out as many wrinkles as possible. Let the visqueen overlap 4 to 10 inches past the forms to accommodate the pulling action that will occur when the stamping tools are pressed into the surface of the concrete.

Using a visqueen reduces the clarity and sharpness of the finish that would ordinarily occur if the tools were applied directly to the surface of the concrete. This produces an antiquing effect. (See Figure 10-2.) If you decide to use the dry cut method of stam-

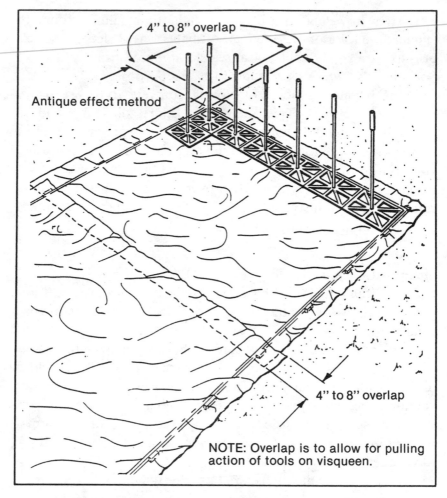

4" to 8" overlap

Antique effect method

4" to 8" overlap

NOTE: Overlap is to allow for pulling action of tools on visqueen.

**Using a visqueen to produce an antique effect
Figure 10-2**

ping (Figure 10-3), you will need the plastic only to cover adjacent structures while pouring. (See Figure 10-4.)

A good rule of thumb is to have one more tool than is needed to cover one width of the area to be stamped. (See Figure 10-5.) Square the tools along the edge of the concrete in an interlocking,

NOTE: Finisher should NOT be more than 2-3 rows ahead of "stampers".

Dry cut method of stamping
Figure 10-3

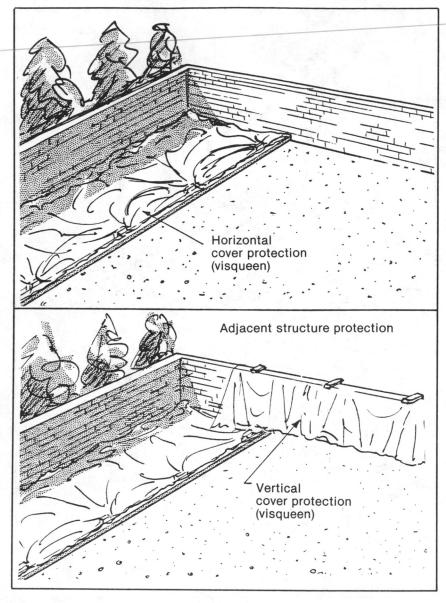

Use plastic to cover adjacent structures while pouring
Figure 10-4

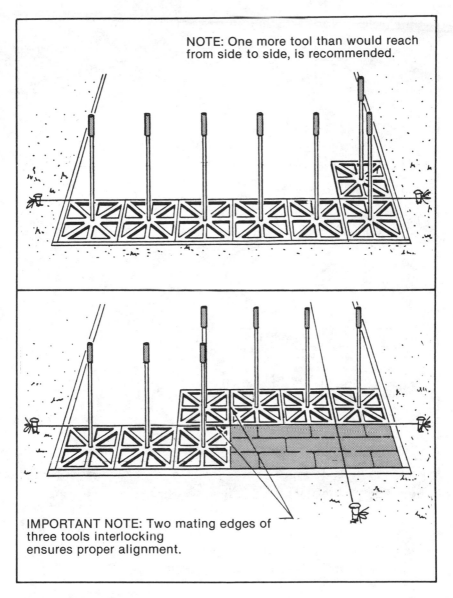

NOTE: One more tool than would reach from side to side, is recommended.

IMPORTANT NOTE: Two mating edges of three tools interlocking ensures proper alignment.

Have one more stamping tool than is needed to cover one width of the area to be stamped
Figure 10-5

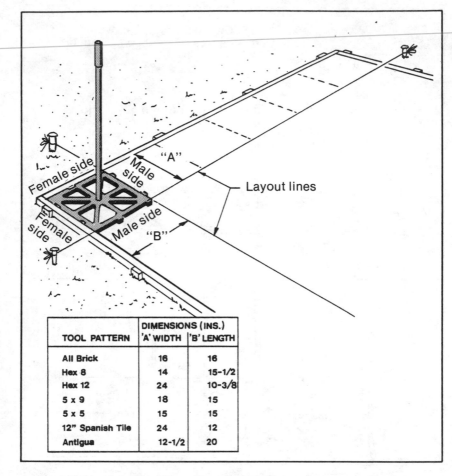

Layout lines

TOOL PATTERN	DIMENSIONS (INS.)	
	'A' WIDTH	'B' LENGTH
All Brick	16	16
Hex 8	14	15-1/2
Hex 12	24	10-3/8
5 x 9	18	15
5 x 5	15	15
12" Spanish Tile	24	12
Antigua	12-1/2	20

**Squaring of tools along the edge to assure proper alignment
Figure 10-6**

straight-line pattern to assure proper placement without misalignment. (See Figure 10-6.) Borders are formed as shown in Figure 10-7.

The concrete should be stamped as soon as it is able to hold its shape. (A common problem is not allowing enough time for stamping.) Tap the stamping tool with a rubber mallet or a hammer

Forming borders with hand tools
Figure 10-7

cushioned by a small 2 by 4 placed on the tool, to form an imprint in the cement. (See Figure 10-8.) Never strike the tool directly with a hammer. As you stamp, make sure you keep the depth of the imprints uniform and even.

Once all the impressions are made, remove the visqueen. When the concrete has hardened to the point that its surface can-

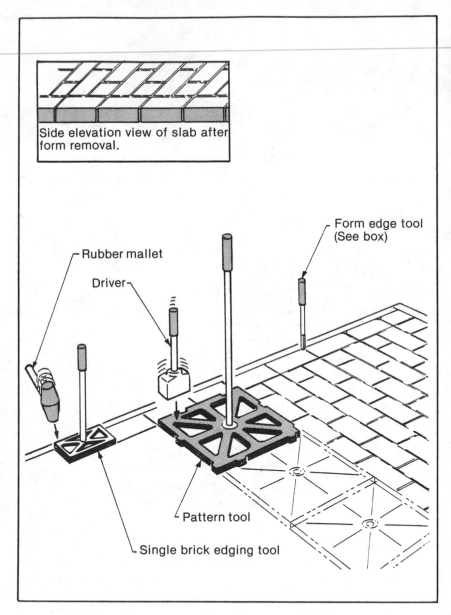

Side elevation view of slab after form removal.

Form edge tool (See box)

Rubber mallet

Driver

Pattern tool

Single brick edging tool

Tap the stamping tool to form an imprint
Figure 10-8

not be marred, apply a spray sealer to promote curing. If colored concrete has been used, a special wax can be applied to add luster and protection to the finish.

Concrete stamping tools are extremely expensive, but well worth the investment. Of the many companies that have created tools for concrete stamping, two stand out. *Indesco-International Design Systems Limited* manufactures the longest-lasting, highest quality stamping tools on the market, and offers a wide range of patterns. *Rafco, Inc.* also produces a variety of high quality stamping tools. These products include a plastic stamping tool used to form various brick patterns on concrete.

11
Sealing, Patching and Repairing

It always pays to take precautions to prevent any problems that might arise. The efficient concrete company always tries to pour in favorable conditions and never takes unnecessary chances.

Sealers
Sealers are a part of preventive maintenance when laying concrete. Use quality sealers liberally as directed on all flatwork, especially exterior work that can be affected by alkalis. An excellent sealer for concrete is *Tri-dar Sealant*. Tri-dar is a linseed oil-based sealer, and if used as directed after pouring, can be invaluable in reducing surface problems, especially in the freeze-thaw areas where flaking can occur.

Sealer can be put down in two ways. The best way to apply it is with a sprayer. Start at any point on the concrete as soon as the surface has hardened to the point where it will not be marred.

Application of a chemical sealant
Figure 11-1

Adjust the spray to get a wide, fine mist. (See Figure 11-1.) Spray in a circular overlapping motion, making sure that the whole surface is sufficiently covered. Properly applied, a gallon of sealer

Characteristic/Type	Mastics	Thermoplastics	Elastomers				One Part Polysulfide and Polyurethanes	Performed Neoprene Compression Gaskets
			Polysulfides	Silicones	Acrylics	Polyurethanes		
Composition	Asphalts, nondrying oils, etc.	Rubber asphalts, Coal tar	Long chain polymers	Silicone polymers	Acrylic acid monomers and polymers-emulsions		Same	Neoprene rubber
Care in surface preparation	Fair	Fair	Critical	Critical	Moderate	Critical to moderate	Same	Joint clean
Cure characteristics	Check manufacturer	Solvent evaporation or by cooling	Catalyst curing agent and 50°F+	Humidity pick up. Temp. not critical	Water evaporation. Temp. not critical	Catalyst curing	Humidity pick up	(Not applicable)
Cure time	When cool	Varies	4 to 7 days			3 to 6 days	PS 20 to 30 days PU 5 to 10 days	(Not applicable)
Handling	Needs equipment. Must be heated	Needs equipment	1 or 2 part 2 part needs equipment	1 part	1 part	1 or 2 part 2 part needs equipment	—	Equipment preferable
Storage	Indefinite	Indefinite	6 months	Limited	Limitless if sealed. Damaged by freezing	Limited	Same	Indefinite
Adhesion	Poor	Fair to excellent	Good	Excellent	Poor	Excellent	PS Good PU Excellent	Function of seal
Low shrinkage	Poor	Poor to excellent	Fair to good	Good	Fair	Good	Same	None
Toxicity—uncured	Moderate	Moderate	Moderate to high	None	Moderate	Moderate	Same	(Not applicable)
Service temperature range	−35°F to 175°F	−35°F to 250°F	−60°F to 250°F	−200°F to 500°F	−35°F to 360°F	−60°F to 250°F	Same	−50°F to 200°F

Sealant characteristics
Table 11-1

Sealant characteristics
Table 11-1 (continued)

Characteristic/Type	Mastics	Thermo-plastics	Elastomers				One Part Polysulfide and Polyurethanes	Performed Neoprene Compression Gaskets
			Polysulfides	Silicones	Acrylics	Polyurethanes		
			75-300 psi.	150-600 psi.	75-150 psi.	150-600 psi.	PS 100-200 psi. PU 300-500 psi.	2000 psi.
Strength—tension	Poor	Poor to excellent	75-300 psi.	150-600 psi.	75-150 psi.	150-600 psi.	PS 100-200 psi. PU 300-500 psi.	2000 psi.
Dynamic properties	Fair to poor	Poor to excellent	Good	Excellent	Fair to poor	Excellent	Same	Excellent
Aging	Fair	Fair to excellent	Fair to good	Good to excellent	Fair	Excellent	Same	Excellent
Weathering	Poor to ultraviolet	Fair to good	Fair—some good	Good	Good	Good	Same	Excellent
Resistance to:								
Water	Excellent	Excellent	Good	Excellent	Excellent	Excellent	Same	Excellent
Solvents (organic)	Poor	Poor	Poor	Fair	Fair	Poor	Same	Excellent
Salts	Excellent	Excellent	Good	Excellent	Excellent	Good	Same	Excellent
Alkalis	Good	Good	Good	Poor	Poor	Good	Same	Excellent
Nonoxidizing acids	Good	Good	Excellent	Good	Excellent	Excellent	Same	Excellent
Oxidizing acids	Good	Good	Poor	Poor	Poor	Poor	Same	Excellent
Recommended uses:								
Highways	No	Yes	Yes	No	No	Yes	Same	Yes
Overpasses and Bridge Decks	No	Yes	Yes	No	No	Yes	Same	Yes
Buildings—metal to metal	No	No (6)	Yes	Yes	Yes	Yes	Same	Yes
Buildings—concrete to joints	No	Yes (1)	Yes, with care (3)	Yes	No	Yes	Same	Yes
Buildings—precast panels	No	No	Yes, with care Yes	Yes	Yes (3)	Yes	Same	Yes
Buildings—caulking	No	No	Yes	Yes	Yes	Yes	Same	Yes
Buildings—floor joints	No	Yes	No	Yes (5)	No	Yes	Same	Yes

Characteristic/Type	Mastics	Thermo-plastics	Elastomers				One Part Polysulfide and Polyurethanes	Performed Neoprene Compression Gaskets
			Polysulfides	Silicones	Acrylics	Polyurethanes		
Underwater—such as:								
Reservoirs	Yes	Yes	Yes (4)	No	No	Yes	Same	Yes
Sewage plants	Yes	Yes	Yes (4)	No	No	Yes	Same	Yes
Water tanks	Yes	Yes	Yes (4)	No	No	Yes	Same	Yes
Basments	Yes	Yes	Yes	No	No	Yes	Same	Yes
Airstrips—Jet resist.	No	Yes, if special (2)	No	No	No	No	Same	Yes
Taxiways—Jet resist.	No	Yes, if special (2)	No	No	No	No	Same	Yes
Runways	No	Yes	Yes	No	No	Yes	Same	Yes
Swimming pools	No	Yes	Yes (4)	No	No	Yes	Same	Yes
Sidewalks	No	Yes	No	Yes (5)	No	Yes	Same	Yes
Recovery	Poor	Excellent	Moderate	100 %	Very poor	100%	PS moderate PU 100%	Excellent
Abrasion and puncture resistance	Poor	Fair	Fair to Poor	Excellent	Poor	Excellent	Same	Excellent
Hardness increase—Age	Poor	Poor to excellent	Poor	Good	Poor	Good	Same	Excellent
Hardness increase—Low temperature	Poor	Poor to good	Poor	Good	Poor	Good	Same	Excellent
Priming	Recommended	Varies, ask manufacturer	Recommended			Recommended	Recommended	Adhesive lubricant

Sealant characteristics
Table 11-1 (continued)

Notes
1. Use only if color is not important.
2. Use only special "jet resistant" thermoplastics.
3. If movement is within capability of sealant.
4. Only if code requirement of exposed area of joint surface per million gallons of water is met—curing agent lead dioxide is toxic.
5. If movement is within sealant capability—also cost is high in large quantities.
6. Thermoplastics have excellent adhesion to metal but they are not recommended here because they cannot be applied with a caulking gun.

will cover about 100 square feet. Keep moving to the side and back while spraying. Wear a pair of shoes with smooth-bottom soles.

If a sprayer is not available, or where a spray mist might ruin surrounding materials inside a building, a roller also gives good results. Use a 4-foot extension handle to make application of the sealer easier.

The sealer has two purposes. First, it helps the concrete retain water, which prolongs curing in the hydration process. This is extremely important because it gives the 28-day curing cycle a good start and if wet curing is maintained for this length of time, a stronger, more durable concrete will result. The second purpose is to protect from dust, chemicals and alkalis by giving the surface of the concrete a protective barrier. At an approximate cost of 2 to 5 cents per square foot of application, sealing is an inexpensive way to ensure durability and protection of concrete construction. Refer to Table 11-1 for a comparison of sealants and their characteristics.

Sealers have an applied effective life span of 300 to 500 freeze-thaw cycles. In many cases, you can reapply the sealer as directed.

Problems such as cracking, chipping and flaking can and do occur. These problems can be repaired through proper procedures. Of course, repairing concrete is not fun, but it must be performed satisfactorily within certain contractual procedures.

Cracking is the most common of all problems with concrete. It can be caused by either expansion and contraction in the normal life span of concrete, or pressure resulting from various sources. Whichever the case, there are many ways to handle crack repairs. Described here are the most economical and practical repair procedures used today.

Repair cracks in concrete flatwork by placing a patching material over the crack. Thoroughly clean the area to be patched to be sure it is free of any loose particles, grease or oil. To patch the crack, use a latex-based material. Mix the patching material with a latex liquid to make a mixture that will adhere to old or new concrete. If properly mixed, its density resembles heavy oil. Pour it onto the crack over an area 1 to 3 inches on each side of the crack. Then trowel the mixture and work it until a uniform blend on the surface has been achieved. A major problem with patching work is the blending in of the patch with the old work. For best results, lightly rub a wet sponge perpendicular to the edge of the patching material and then broom if the surface requires a broomed finish. The latex-based material can be used for all flatwork cracks. It is strong, durable and, for major repairs, very economical.

Patching residential concrete foundation walls requires a lasting watertight patch of a material far stronger than latex. This is an epoxy-type patching material. An epoxy patch is mixed much like a glue is mixed. Apply it either by trowel or by flat scraper to the crack and 3 to 6 inches on either side of the crack. Make sure the surface is dry and rough.

Epoxy patching will not look like a factory job, but it can correct many defects that may occur in concrete walls. When patching with epoxy, don't be afraid to use it liberally because the strength is in the amount used.

In some instances, especially with flatwork, cracks become too wide and uneven to repair. The best way to repair this type of potentially dangerous problem is to replace the flatwork.

If you place a good quality concrete with proper preparation, finishing and curing, the chances of developing concrete problems will be minimal. Remember, it is always cheaper to do your best on the job from the beginning than to have to spend extra time and money to repair concrete resulting from poor workmanship and placement.

12
Estimating

T he subject of estimating is of utmost importance because it is in this critical area that a business venture can prosper or fail. The viability of a contracting business is in essence controlled by the accuracy of its estimators. Only with a careful, proven system of estimating and constant close monitoring of the daily operations within the company can profitability and prosperity continue.

There are many constants and variables that will affect the bid proposal. The size of a job is the first variable that you should consider. It is essential to know if a competitive bid can be rendered, because there is no point in taking on a job that you can't handle. In other words, make sure the company can perform the work bid, considering all the difficulties of the job. Too often in the bidding, a close look at the jobsite is not made and special conditions the job may present are overlooked. Failure to observe and understand all aspects of the job can certainly affect that thin line separating a profit from a loss. Estimating how long it takes to perform specific tasks takes experience. Once this experience is gained, more accurate estimates will follow.

Most contractors formulate their answers on one tabulation sheet to come up with an estimate. Presented here is a unique and accurate way to estimate any job.

The three variables affecting an estimate are materials, labor and equipment. In figuring a particular job, you must first know what materials are needed and their costs. An extremely important information sheet called an *Abstract of Bids Sheet* lists the major competitive material suppliers. From this data sheet, you can see how the material prices fluctuate by day, month or year. These changes are extremely important. Another listing on this sheet is the time it takes to deliver the required materials. The sheet makes it possible to have current prices available at one's fingertips. A sample of this bid sheet is shown in Figure 12-1.

Another important chart is a *Production Rate Listing*. On this sheet, the people required for a particular job bid are listed, allowing you to estimate a total labor cost per day. As an example, consider a 5,000 linear-foot M-3-12 curb and gutter job (mountable type with a 3-inch high roll and a 12-inch high back). From experience, I estimate that for any mountable or rolling curb, the production rate is 75 linear feet per man per day. An average curb crew consists of five to six men. The labor is broken down on this chart in the following fashion. The chart lists one foreman, two laborers, one finisher and one operator. Each man earns an established union scale per hour. When rates are marked down, each man's labor costs can be established by the number of hours worked. Assume a 5,000-foot curb job with each man on a 5-man crew doing 75 feet per day. The total crew production each day is 375 feet. Divide 375 into 5,000: the five men will take approximately 14 days to complete this job. Refer to Figure 12-2.

Abstract of Bids
Middlewest Development, Inc.

Bid To _____
Location _____

Sheet No. ____ of ____ Sheet

Bidder _____
Person Contacted _____
Date _____
Telephone No. _____

Item No.	Material	Quantity	Unit	Unit Price	Total Price	Unit Price	Total Price	Unit Price	Total Price	Unit Price	Total Price
1											
2											
3											
4											
5											
6											
7											
8											
9											
10											
11											
12											
13											
14											
15											
16											
17	Summary Total Price										
18	Promised Delivery and F.O.B. Point										

Sample bid sheet
Figure 12-1

Class	No.	Rate	8 Hour	9 Hour	10 Hour
Foreman	1				
Laborers	2				
Finishers	1				
Operators	1				
Totals					

Production Rates

Production rate listing
Figure 12-2

Each job involves different equipment. With an equipment chart (Figure 12-3), you can list the equipment owned, and by using the formula of the list price, including interest, times 1.4 divided by 3, divided again by 1,200 hours, you can calculate the cost of operation of the equipment over a span of three years. The 1,200-hour figure is assuming that a construction company works seven months a year, 40 hours per week. Total the equipment costs and you have a cost-per-day figure for the curb job. Post this figure on an *Item Cost Analysis Chart* as shown in Figure 12-4. Record owner, project, location and job-type data which will formulate the cost totals by item. By referring to the equipment cost, production rate and material costs, then transferring the prerecorded total to the Item Cost Analysis Chart, you can see ac-

Chart For Curb Job

Equipment	Hrs.	Hour Rate	Curb	Remove & Repour	Flat Work
EQUIPMENT					
Pick-up	8	$3.50	$28.00		
Form Truck	8	5.00	40.00		
Case	8	7.50	60.00		
Trailer	8	2.00	32.00		
Dump	8	6.10	52.00		
Concrete saw	8	1.50			
Pump	8	5.00			
Trowel Machine		2.00			
TOTAL			$212.00		

Per Day

Each particular job involves different equipment usage.

Your cost per day is $212.00

Equipment chart for curb job
Figure 12-3

curately what the cost will be for the 5,000 feet of M-3-12 curb and gutter per linear foot. Now transfer the $4.51 unit cost total to a *Unit Price Bid Summary Sheet* shown in Figure 12-5. Keep in mind that $4.51 is your cost per linear foot of curb and gutter poured.

To this figure, add the profit needed to complete the curb job. Overhead, or operating costs, is governed by many variables, each contractor having unique costs. To help determine the profit that

				SHEET OF ESTIMATE NO. JOB NO.	

ITEM COST ANALYSIS

OWNER

PROJECT

LOCATION

RTE. & SEC. NO.

ITEM NO. CODE	Quantity	Unit	COST			
			SUBC'T	LABOR	EQUIPMENT	MATERIAL
M- 3 -12 CURB & GUTTER 10"FL	5000	LF				46. 00
SAMPLE EXAMPLE						per yard
CONCRETE 10,580 —	230 Cy	CY				
LABOR 6,760 —	13	DAYS		520 00		
EQUIPMENT 2,756 —	13	DAYS			212 00	
HARDWARE - 50. 0 C. 500 —	100	EACH				5 00
FORMS. 750 —	5000	LF				
STONE - 650 —	100	T				6 20
TAX - IF (APPLICABLE)						
ON MATERIALS 22,585.57						
DIVIDE 5000 INTO 22,585.57						
AND YOU GET A UNIT COST OF						
$4.51 PER LINEAL FOOT OF						
CURB & GUTTER POURED						
HARDWARE IS (EXPANSION) (NAILS)						
DOWEL BARS SLEEVES ETC						
TOTALS						

Item cost analysis chart
Figure 12-4

		UNIT PRICE BID SUMMARY					SHEET OF ESTIMATE NO. JOB NO.			

OWNER		ARCH'T. ENG.			
PROJECT		BID DUE - DATE		HOUR	
LOCATION		TIME ALLOWANCE		CALENDAR DAYS WORKING DAYS	
RTE. & SEC. NO.		PENALTY		BONUS	
LETTING BY		BID DEPOSIT			

#	CODE	ITEMS OF WORK	QUANTITY	Unit	COST Unit		COST Amount		BID Unit		BID Amount	
1		M - 3-12 CURB + GUTTER 10"FL	5000	LF	4	51	22	580 00	6	01	30 050

NOTE.
COST UNIT $\perp$ .75 = 6.01
A 25% MARK UP

TOTAL FIELD COST				
ADJUSTMENTS				
GENERAL CONDITIONS				
PROFIT & G.O. OVERHEAD				
BONDS				
BONDS DISCOUNT				
TOTAL BID AMOUNT				

Unit price bid summary sheet
Figure 12-5

you can live with, use the following formula:

Profit equals Total Overhead plus the desirable Profit.

It is a good idea to acquire an individual overhead breakdown of the company. The average markup is between 19 and 40 percent. Now take the unit cost of $4.51, divide by 75 (which is a 25 percent markup) and the total bid figure that would be submitted on the unit price bid summary is $6.01. See sample in Figure 12-5.

Remember what your aims should be in estimating. First, you want to be correct in the estimate; second, you want to be the low bidder. With a little carelessness you can always be the low bidder. But you want to be correct, *and* the low bidder, which is difficult. It means figuring your quantities accurately, using the lowest material prices possible, and analyzing your labor units to keep them as low as possible. Every amount that goes into the money columns in an estimate should be reviewed for possible savings. This is the only way to compete with the consistently careless low bidders.

Don't forget that you have made an estimate when you are awarded a job. Your labor costs should go to the foreman or superintendent on the job in a form he can understand. This does not necessarily mean in dollar figures. Perhaps days of working time would be easier for your operation. Or, perhaps with a certain size crew, the foreman can be told how far the job should have progressed at the end of each week. However this is done, it must be set up by the office and checked constantly. Waiting until the job is 50% complete to find out that you have spent 75% of the money figured for labor is too late. The time to correct a problem is when the job is 10% complete and running 20% of the money figured. If this happens and the estimate is correct, the foreman or superintendent carries part of the blame for the higher costs, but it is the responsibility of management to bring costs back into line.

Estimating Tables

Materials for Concrete

Sacks of Cement	Gallons of Water Per Sack of Cement	Weights of Saturated Surface-Dry Aggregate Per Sack of Cement, Lbs.		28-Day Compressive Strength P.S.I.
		Sand	Gravel	
4.9	8	280	380	2,250
5.9	7	240	330	2,750
6.0	6½	210	310	3,000
6.5	6	190	280	3,300
7.2	5½	160	260	3,700
8.0	5	140	230	4,250

Maximum Size of Aggregate Recommended

Minimum Dimension of Section (Inches)	Maximum Size of Aggregate* in Inches, for:			
	Reinforced Walls, Beams, and Columns	Unreinforced Walls	Heavily Reinforced Slabs	Lightly Reinforced or Unreinforced Slabs
2½– 5	½– ¾	¾	¾–1	¾–1½
6 –11	¾–1½	1½	1½	1½–3
12 –29	1½–3	3	1½–3	3
30 or more	1½–3	6	1½–3	3 –6

*Based on square openings.

Square Feet of Concrete From
1 Cubic Yard of Concrete

Thickness Inches	S.F.	Thickness Inches	S.F.	Thickness Inches	S.F.	Thickness Inches	S.F.
1	324	4	81	7	46	10	32
1¼	259	4¼	76	7¼	44	10¼	31
1½	216	4½	72	7½	43	10½	31
1¾	185	4¾	68	7¾	42	10¾	30
2	162	5	65	8	40	11	29½
2¼	144	5¼	62	8¼	39	11¼	29
2½	130	5½	59	8½	38	11½	28
2¾	118	5¾	56	8¾	37	11¾	27½
3	108	6	54	9	36	12	27
3¼	100	6¼	52	9¼	35	12¼	26½
3½	93	6½	50	9½	34	12½	26
3¾	86	6¾	48	9¾	33	12¾	25½

Thickened edges must be calculated separately

Slabs
Cubic Measure per 100 Square Feet

	Per 100 Square Feet Slab	
Thickness	Cubic Feet of Concrete	Cubic Yards of Concrete
2″	16.7	.62
3″	25.0	.93
4″	33.3	1.24
5″	41.7	1.55
6″	50.0	1.85

Labor and Materials for 100 S.F. of Foundation Wall Forms

Work Element	B.F. per S.F. of Forms	Make and Place Forms			Removing Forms	
		S.F. in 8 Hours	Carpenter Hours	Labor Hours	S.F. in 8 Hours	Labor Hours
Foundation wall forms						
4' to 6'	2	190-210	4.0	2.0	640	1.3
7' to 8'	2	165-190	4.5	2.3	550	1.5
9' to 10'	2.5	150-160	5.3	2.5	450	1.8
11' to 12'	3	135-150	5.5	2.8	425	1.9
Retaining walls						
16' to 20'	3.5	105-115	7.3	3.5	325	2.5

Conversions for Reinforcing Bars

Bar Number	Lbs. Per L.F.	Diameter in Inches	Cross-Sectional Area, Inches	Perimeter in Inches	L.F. Per Ton
3	.376	0.375	0.11	1.178	5,319
4	.668	0.500	0.20	1.571	2,994
5	1.043	0.625	0.31	1.963	1,918
6	1.502	0.750	0.44	2.356	1,332
7	2.044	0.875	0.60	2.749	978
8	2.670	1.000	0.79	3.142	749
9	3.400	1.128	1.00	3.544	588
10	4.303	1.270	1.27	3.990	465
11	5.313	1.410	1.56	4.430	376
14	7.650	1.693	2.25	5.320	261
18	13.600	2.257	4.00	7.090	147

The nominal dimensions of a deformed bar are equivalent to those of a plain round bar having the same weight per foot as the deformed bar.

Bar numbers are based on the number of eighths of an inch included in the nominal diameter of the bars.

Labor and Materials for Excavating and Pouring Footings, Foundations and Grade Beams

	Material			Labor	
Size	C.F. Concrete Per L.F.	C.F. Concrete Per 100 L.F.	C.Y. Concrete Per 100 L.F.	Man-Hours Per 100 L.F. for Hand Excavation	Man-Hours Per C.Y. for Concrete Placement
6 x 12	0.50	50.00	1.9	3.8	1.1
8 x 12	0.67	66.67	2.5	5.0	1.1
8 x 16	0.89	88.89	3.3	6.4	1.1
8 x 18	1.00	100.00	3.7	7.2	1.1
10 x 12	0.83	83.33	3.1	6.1	1.0
10 x 16	1.11	111.11	4.1	8.1	1.0
10 x 18	1.25	125.00	4.6	9.1	1.0
12 x 12	1.00	100.00	3.7	7.2	1.0
12 x 16	1.33	133.33	4.9	9.8	1.0
12 x 20	1.67	166.67	6.1	12.1	0.9
12 x 24	2.00	200.00	7.4	15.8	0.9

Reduce excavation hours by ¼ for sand or loam. Increase excavation hours by ¼ for heavy clay soil.

Placement labor is based on ready-mix concrete direct from chute. Add 50% if concrete is pumped into place. Add 60% if concrete is placed with a crane and bucket. Add two hours per cubic yard if concrete is wheeled 40 feet into place and an additional 25% for each additional 40 feet.

Excavation includes loosening and one throw from trench only.

Index

Practical References For Builders

Masonry & Concrete Construction
Every aspect of masonry construction is covered, from laying out the building with a transit to constructing chimneys and fireplaces. Explains footing construction, building foundations, laying out a block wall, reinforcing masonry, pouring slabs and sidewalks, coloring concrete, selecting and maintaining forms, using the Jahn Forming System and steel ply forms, and much more. Everything is clearly explained with dozens of photos, illustrations, charts and tables.
224 pages, 8½ x 11, $13.50

Estimating Home Building Costs
How to estimate every phase of residential construction from site costs to the profit margin you should include in your bid. Shows how to keep track of manhours and make accurate labor cost estimates for footings, foundations, framing and sheathing finishes, electrical, plumbing and more. Explains the work being estimated and provides sample cost estimate worksheets with complete instructions for each job phase. Includes tables, charts, and illustrations to help you accurately estimate home building costs.
320 pages, 5½ x 8½, $14.00

Construction Estimating Reference Data
Collected in this single volume are the building estimator's 300 most useful estimating tables. Labor requirements for nearly every type of construction are included: site work, concrete work, masonry, steel, carpentry, thermal & moisture protection, doors and windows, finishes, mechanical and electrical. Each section explains in detail the work being estimated and gives the appropriate crew size and equipment needed. Many pages of illustrations, estimating pointers and explanations of the work being estimated are also included. This is an essential reference for every professional construction estimator.
368 pages, 11 x 8½, $18.00

Concrete Construction & Estimating
Explains how to estimate the quantity of labor and materials needed, plan the job, erect fiberglass, steel, or prefabricated forms, install shores and scaffolding, handle the concrete into place, set joints, finish and cure the concrete. Every builder who works with concrete should have the reference data, cost estimates, and examples in this practical reference.
571 pages, 5½ x 8½, $15.75

National Construction Estimator
Current building costs in dollars and cents for residential, commercial and industrial construction. Prices for every commonly used building material, the proper labor cost associated with installation of the material. Everything figured out to give you the "in place" cost in seconds. Many time-saving rules of thumb, waste and coverage factors and estimating tables are included. **288 pages, 8½ x 11, $10.75. Revised annually. Also available in 42X Microfiche for $5. Monthly cost updates for 280 key materials and the wages in the National Construction Estimator are available for $2 per month. Order the "National Construction Cost Newsletter" for $24 per year.**

Reducing Home Building Costs

Explains where significant cost savings are possible and shows how to take advantage of these opportunities. Six chapters show how to reduce foundation, floor, exterior wall, roof, interior and finishing costs. Three chapters show effective ways to avoid problems usually associated with bad weather at the jobsite. Explains how to increase labor productivity. **224 pages, 8½ x 11, $10.25**

Building Layout

Shows how to use a transit to locate the building on the lot correctly, plan proper grades with minimum excavation, find utility lines and easements, establish correct elevations, lay out accurate foundations and set correct floor heights. Explains planning sewer connections, leveling a foundation out of level, using a story pole and batter boards, working on steep sites, and minimizing excavation costs. **240 pages, 5½ x 8½, $11.75**

Basic Plumbing With Illustrations

The journeyman's and apprentice's guide to installing plumbing, piping and fixtures in residential and light commercial buildings: how to select the right materials, lay out the job and do professional quality plumbing work. Explains the use of essential tools and materials, how to make repairs, maintain plumbing systems, install fixtures and add to existing systems. **320 pages, 8½ x 11, $15.50**